SEMIORDERS

THEORY AND DECISION LIBRARY

General Editors: W. Leinfellner (*Vienna*) and G. Eberlein (*Munich*)

Series A: Philosophy and Methodology of the Social Sciences

Series B: Mathematical and Statistical Methods

Series C: Game Theory, Mathematical Programming and Operations Research

Series D: System Theory, Knowledge Engineering and Problem Solving

SERIES B: MATHEMATICAL AND STATISTICAL METHODS

VOLUME 36

Scope: The series focuses on the application of methods and ideas of logic, mathematics and statistics to the social sciences. In particular, formal treatment of social phenomena, the analysis of decision making, information theory and problems of inference will be central themes of this part of the library. Besides theoretical results, empirical investigations and the testing of theoretical models of real world problems will be subjects of interest. In addition to emphasizing interdisciplinary communication, the series will seek to support the rapid dissemination of recent results.

SEMIORDERS

Properties, Representations, Applications

by

M. PIRLOT

Faculté Polytechnique de Mons,
Mons, Belgium

and

PH. VINCKE

Université Libre de Bruxelles,
Bruxelles, Belgium

KLUWER ACADEMIC PUBLISHERS
DORDRECHT / BOSTON / LONDON

A C.I.P. Catalogue record for this book is available from the Library of Congress.

ISBN 978-90-481-4869-1

Published by Kluwer Academic Publishers,
P.O. Box 17, 3300 AA Dordrecht, The Netherlands.

Sold and distributed in the U.S.A. and Canada
by Kluwer Academic Publishers,
101 Philip Drive, Norwell, MA 02061, U.S.A.

In all other countries, sold and distributed
by Kluwer Academic Publishers Group,
P.O. Box 322, 3300 AH Dordrecht, The Netherlands.

Printed on acid-free paper

To our wives
Nuran and Annie
and our children
Yunus Can,
Lionel and Johanne

Contents

INTRODUCTION

Semiorder is probably one of the most frequently used ordered structure in all areas of Science. Scientists build models, representations of what they observe. The iterative process of model building and checking relies upon endless comparisons: between several observations, between results of different models, or between observations and results. These comparisons can be realized via human perceptions or via very sophisticated and precise instruments. In any case, there always exists a threshold under which differences are not perceptible any more. Thus, observations or results which are very slightly different will be declared "equal". This will inevitably lead to situations where a is "equal" to b, b is "equal" to c but a is not "equal" to c. This fact has been acknowledged in the past by scientists like G. Fechner as early as in 1860, by H. Poincaré in 1905 and by others (excerpts from their texts have been reproduced in chapter 2). In the area of decision-analysis and preference modelling, which is the field of activity of the authors of this book, it is R. D. Luce who pointed out, in 1956, the phenomenon of intransitivity of "equality". It must be noted that this intransitivity is in contradiction with the traditional model used by the scientists who build decision-aid tools, in fields like operations research, statistics, economy, finance or insurance. In that traditional model, preferences are represented by a real-valued function g, defined on the set of possible decisions (feasible solutions), such that decision a is preferred to b if the value $g(a)$ is greater (for example) than $g(b)$. Finding the best decision is then possible by maximizing function g (which can be a very difficult task from a technical point of view). With such a model, there is no preference between a and b (they are "equal") if and only if $g(a) = g(b)$. As equality is transitive in the set of real numbers, this shows that the model does not correctly account for the above mentioned phenomenon (a is "equal" to b and b is "equal" to $c \Rightarrow g(a) = g(b)$ and $g(b) = g(c) \Rightarrow g(a) = g(c) \Rightarrow a$ is "equal" to c). In recent years, several decision-aid methods have been developed, especially in the field of multicriteria decision-aid, in order to explicitly take into account the intransitivity of "equality". In these methods, there is no preference between a and b (they are "equal") if the difference between $g(a)$ and $g(b)$ is, in absolute value, less than a certain threshold. Of course, this is a generalization of the traditional model which corresponds to the case where the threshold is null. The preference structure defined by this new model is called *semiorder*. The universality of the above mentioned phenomenon (intransitivity of "equality") led us to consider *semiorder* as a very fundamental structure; we decided to dissect it and to try to reveal some of its hidden treasures. One of these treasures is the notion of minimal representation which provides a unique "natural" numerical representation of the semiorder (in the same way as numbering provides a "natural" numerical representation of a ranking): this no-

1

tion is developed in chapter 4. Another interesting aspect (developed in chapter 6) is the possibility, on the basis of the concept of semiorder, to build a general theoretical framework for multicriteria decision-aid methods. Besides these two important aspects, we also propose in this book a chapter showing how semiorders appear in a lot of different fields (chapter 2) and a chapter on the generalization of this concept to valued (or fuzzy) relations (chapter 5).

Each of the chapters 4, 5, 6 can be read (almost) independently provided the material contained in chapter 3 is familiar to the reader. We hope that chapters 1 and 2 can help to stimulate the interest of those who are mainly concerned with applications. This book is certainly not an exhaustive survey of all what can be said about semiorders (as shown in chapter 7) but we hope that the reader will be convinced of the interest of this structure and will be encouraged to teach it and to use it in decision-aid methods as well as in other modelling contexts. In contrast with Fishburn's book on interval orders and interval graphs (1985), a subject very close to ours, the present work could be described as less order-theoretic and more applications oriented. In our main chapters (4,5 and 6), we concentrate on questions that are of interest to students (at graduate level) and researchers in the fields of decision analysis, management science, operations research, discrete mathematics, classification, social choice theory, order theory and to practitioners designing decision tools and methods.

Semiorders: where can they arise from?

Notwithstanding the descriptions of applications of semiorders in various fields, which can be found in chapter 2, we want to show here that semiorders can arise as a natural model for several very general situations. Consider a finite set A of objects with a number $g(a)$ attached to each object $a \in A$. This number may for instance result from a measure performed on each object; to some extent, such a measure is always imprecise, as already observed. When imprecision may be considered constant within the range where the measurement apparatus is used, let δ denote the value of this constant; the precise measure of object a is between $g(a) - \delta$ and $g(a) + \delta$ and the ordered structure on the objects, induced by the measurement, is a semiorder. Indeed, it may be asserted that object a is larger than object b (w.r.t. the measured characteristic) only if the observed measure of a passes the observed measure of b by at least twice the constant imprecision threshold of the apparatus. Moreover, the observed measure g of each object, together with twice the imprecision threshold, constitute a constant threshold representation of the semiorder on the set of objects A, i.e. a very particular and quite specific numerical representation of the semiorder. In general, the imprecision is not constant along the measurement scale. The measure of object a lies between $g(a) - \delta(a)$ and $g(a) + \delta(a)$ where $\delta(a)$ is the imprecision on the observed value of a and depends on a; a will be declared larger than object b only if the most pessimistic value for the measure of a, $g(a) - \delta(a)$, passes the most optimistic evaluation of b, $g(b) + \delta(b)$. However, the resulting structure on A is still a semiorder as soon as the function δ is such that no interval $[g(a) - \delta(a) , g(a) + \delta(a)]$

is included in any interval $[g(b) - \delta(b) \, , \, g(b) + \delta(b)]$ or alternatively, as soon as the order of the lefthand endpoints of the intervals $[g(a) - \delta(a) \, , \, g(a) + \delta(a)]$ is the same as the order of the righthand endpoint of those intervals. It is so, for instance, when the imprecision threshold is constant in relative value, i.e. when $\delta(a)$ is proportional to $g(a)$. From the above, we observe that a set of semiordered objects can either be viewed as represented by a function and a threshold (which is variable, in general) or by intervals of the real line; semiorders are particular cases of interval orders. In preference modelling and in other contexts as for instance, the comparison of stimuli in sensory experiments, an interval representation of the objects is not the most natural. For instance, when comparing noise intensities, a difference is perceived as soon as the ratio of the stimuli intensities passes some threshold; in terms of difference of intensities, the threshold is proportional to the smaller stimulus intensity. Taking the logarithm of the intensities, leads to a model which is formally the same as above; the semiorder on the stimuli admits a numerical representation with constant threshold, where each stimulus is represented by the logarithm of its intensity and the "just noticeable difference" is the logarithm of the threshold on the ratios. The decibel scale for noise or sound comparison, is the most famous example of such a model. In the above situations, semiorders have arisen through a particular and natural numerical representation (with threshold). This is not always the case; a semiorder can also be given as a binary relation without numbers attached to the objects. This may occur for instance when the numerical representation has been lost or has no meaning beside the binary relation it induces or when the semiorder is obtained through a procedure like preference aggregation performed on binary relations and yielding a binary relation. In that case, many different numerical representations can be provided including infinitely many constant threshold representations and infinitely many variable threshold representations. Special representations like integer valued ones are also of interest. It should be stressed that the choice of a particular numerical representation is in general arbitrary, unless additional information is available for selecting it. Semiorders or more precisely their symmetric part called "indifference graphs", also emerge as pertinent structures for representing similarities among objects. Starting from numbers attached to the objects and interpreting the fact that a pair of these numbers are "not too different" as indicating that the corresponding objects are similar, yields an indifference graph on the set of objects. This structure is the simplest model for nontransitive similarities. An interesting fact is that an indifference graph almost uniquely determines a semiorder; there are only two semiorders whose symmetric part is a given indifference graph and they just differ by the orientation. This means that the study of indifference graphs almost reduces to the study of semiorders or conversely. In this work we concentrate on the ordered structure, but of course, many conclusions can be transposed into the language of indifference graphs.

Contents of the book

The first chapter aims at giving the reader the basic notions which are necessary for understanding the applications described in chapter 2; definition, matricial and graphical representations of a relation, basic definitions of semiorder, interval order and of their non- oriented associated relations, introduction to the valued semiorder structure. The applications described in chapter 2 are concerned with a variety of fields such as genetics, archaeology, operations research, computer science, decision-aid, classification,.... The rigorous presentation of the semiorder structure and its properties is the subject of chapter 3, where several equivalent definitions are presented and where the distinction is made between a semiorder and a strict semiorder. More particular structures, like the complete preorder and order and more general ones, like the interval order are also described. In general, the properties which are mentioned in this chapter are known in the literature; we nevertheless give the proofs so that the reader has an unified presentation of the main results. A particularity of the semiorder is that it is, among the structures which generalize the classical notion of ranking (linear order), the one closest to a linear order. Indeed, to every semiorder on a set, corresponds univocally a strict complete order on this set. A natural question is then to know whether it is possible to number the elements of a semiordered set in a way that is similar to the numbering of an ordered set by the integers $1, 2, 3, \ldots$. The answer to this question resides in the concept of minimal representation, studied in chapter 4. Associating a particular valued graph to any semiorder and using the notion of potential function on this valued graph, allows to derive a necessary and sufficient condition for the existence of a numerical representation of a given semiorder, with a given threshold q; an algorithm is provided to build that numerical representation. The minimal representation is the particular representation which simultaneously minimizes the values representing all the elements of the semiordered set; we prove that every semiorder has a minimal representation (this was not obvious a priori), that this representation exists in the set of integers and that it is the most "contrasting" among all the representations in a given interval. We then study two particular relations associated to a semiorder, which are the sets of so-called noses and hollows: these relations allow to find new and very economical representations of a semiorder, by graphs which are called synthetic and super-synthetic. We also present in this fourth chapter the minimal representation of the interval orders. Chapter 5 is devoted to the valued semiorders. Valued relations arise in fields like psychological studies on preference or discrimination, classification and decision-aid. Given a set A of elements $a, b, \ldots$, a value $v(a, b)$ is associated to each pair (a, b), representing either the proportion of times a given subject judges stimulus a to be "greater" than stimulus b either the proportion of individuals who prefer a to b, either the "similarity" between a and b or the credibility or intensity of preference of a over b. A very abundant literature is devoted to valued relations in mathematical psychology and in fuzzy sets theory (where "valued" is replaced by "fuzzy"). A valued semiorder is a valued relation such that every relation obtained by fixing a threshold α and by keeping the pairs of elements which have a value greater or equal to α, is a semiorder (or a strict semiorder).

When α varies, one obtains an imbedded family of semiorders and strict semiorders; in chapter 5, we study the numerical and minimal representations of these structures. An interesting particular case concerns a family of semiorders which have the same underlying complete order. This leads to a structure which can be represented by one real function and several thresholds (which can correspond to several degrees of preference). We also consider the case where these thresholds are all constants. The aggregation of semiorders is analyzed in chapter 6. It is remarkable that passing from complete orders to semiorders completely changes the properties of classical aggregation procedures; we illustrate that point on the lexicographic method and Borda's method and we propose new versions of these methods, adapted to the case of semiorders. However, although the set of semiorders is larger and hence offers more freedom for modelling, it turns out that there is not enough flexibility in it to avoid the difficulties encountered with complete orders; in particular, wanting to aggregate any family of complete orders into a semiorder, we are faced with an impossibility result of Arrow-type. Note that it is even the case if we try to aggregate into a still more relaxed structure like interval orders.

In the last part of the chapter, we explore the paths that could be followed to develop a theory of semiorders aggregation that would be comprehensive enough to encompass most existing schemes for aggregating preference structures. There are two main ideas underlying most models of preference aggregation. In the most classical one, it appears that the overall preference can be described as the result of an operation on numerical representations of the component structures. This yields what can be interpreted as an overall evaluation of the objects (points, alternatives) on a single other "dimension" that synthesizes the various "component dimensions". Alternatively, and this is also a generalization, the global structure can be viewed as arising from the pairwise comparison of the objets, through a combination of the discrepancies between the two objects on each component dimension. Usually with the second approach, the aggregated structure lacks strong coherence properties such as transitivity; this is little wonder since the overall relation between two objects is determined for these objects independently of the others. Belonging to the first category are the multiattribute utility or value models; to the second, some of the outranking models, including particular versions of ELECTRE. What we do essentially is to present characterizations of the overall structures that could be obtained in either of the two modes, starting from components that are semiorders. The results extend to interval orders; they also extend to valued semiorders which allows to encompass outranking models such as valued versions of ELECTRE or PROMETHEE. The final chapter touches on some topics which are not covered in this book; some bibliographic indications are provided. The main limitation in the scope of the present work is that only semiorders on finite sets of objects are dealt with. We say one word about the problems raised by semiorders on infinite sets and refer the interested reader to the literature. Then, we go through a number of themes like semiorders on mixture sets (in connexion with decision making under risk), partial semiorders, dimension, enumeration problems about semiorders, algorithmic aspects and the study of indifference graphs.

6

Acknowledgments

It is a great pleasure to acknowledge the contributions of our friends Denis Bouyssou, Thierry Marchant, Patrice Perny and Alexis Tsoukias, with whom we work in close cooperation for many years on problems of decision analysis and preference modelling. We are grateful to them for reading the manuscript and formulating many remarks that helped us correct mistakes and improve the presentation of the results. We are also indebted to Jean-Paul Doignon; he generously let us benefit of his knowledge and views on the field. We thank Bernard Roy for his encouragements, his permanent interest and a number of discussions on topics dealt with in this book.

Finally, special thanks are due to Berthold Ulungu for his patience in typing the various versions of the manuscript in LaTeX as well as to Philippe Fortemps for his skill in solving all kind of delicate problems encountered when trying to obtain what we wanted from the LaTeX system.

1

FIRST PRESENTATION OF THE BASIC CONCEPTS

Preliminary remark

The purpose of this chapter is not to give the formal definitions of the concepts which will be used in this book (these formal definitions will be given in chapter 3). We just want to introduce here the vocabulary needed to understand chapter 2, in which we expose the motivations for studying semiorders. We have avoided, as much as possible, the mathematical developments, so that the reader who is mainly interested in the applications can understand chapters 1 and 2 without too much effort.

1.1 Basic definitions and notations

This book is devoted to the study of certain types of relations which are defined on a finite set. In this section, we recall the basic vocabulary about relations.

We consider a finite set denoted by A; its elements are generally denoted $a, b, c, \ldots$, and sometimes $x, y, z, \ldots$.

A binary relation S on the set A is a set of ordered pairs denoted (a, b), where a and b are elements of A. If (a, b) belongs to S, we write

$$(a, b) \in S \text{ or } aSb;$$

if not,

$$(a, b) \notin S \text{ or } a\neg Sb.$$

A binary relation S on the set A is

- *reflexive* iff aSa, for all $a \in A$,

- *symmetric* iff $aSb \Rightarrow bSa$, for all $a, b \in A$,

- *asymmetric* iff $aSb \Rightarrow b\neg Sa$, for all $a, b \in A$,

- *complete* iff aSb or bSa, for all $a, b \in A$, $a \neq b$,

- *transitive* iff aSb and $bSc \Rightarrow aSc$, for all $a, b, c \in A$.

Given a binary relation S on the set A, we respectively denote P_s and I_s the *asymmetric* and the *symmetric parts* of S:

$$aP_sb \text{ iff } aSb \text{ and } b\neg Sa,$$

$$aI_sb \text{ iff } aSb \text{ and } bSa.$$

It is clear that

$$aSb \Rightarrow aP_sb \text{ or } aI_sb \text{ (exclusive ``or'')}.$$

Given a binary relation S on the set A, two elements a and b are said to be *equivalent* for S iff, for all $c \in A$,

$$\left\{ \begin{array}{lll} aSc & \text{iff} & bSc, \\ cSa & \text{iff} & cSb. \end{array} \right.$$

A binary relation S on a set A can be represented by a directed graph whose set of nodes is A and where there exists an arc from a to b iff aSb. When S is symmetric, the two arcs from a to b and from b to a can be replaced by an edge between a and b, giving an undirected graph.

Another way for representing a binary relation S on a set A is to associate, to each element of A, a line and a column of a matrix M^s. The element M^s_{ab} of this matrix (at the intersection of the line associated to a and the column associated to b) is equal to 1 if aSb and equal to zero if not.

A valued binary relation v on the set A is a mapping from $A \times A$ to the set of real numbers $\mathcal{R}$; to each ordered pair (a, b) of elements of A, is associated a real number $v(a, b)$. A valued binary relation can be represented by a valued directed graph where the value $v(a, b)$ is associated to the arc from a to b. It can also be represented by a matrix M^v where

$$M^v_{ab} = v(a, b), \text{ for all } a, b \in A.$$

Figures 1.1 and 1.2 illustrate the previous definitions.

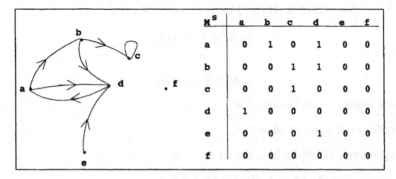

M^s	a	b	c	d	e	f
a	0	1	0	1	0	0
b	0	0	1	1	0	0
c	0	0	1	0	0	0
d	1	0	0	0	0	0
e	0	0	0	1	0	0
f	0	0	0	0	0	0

Figure 1.1: Example of binary relation

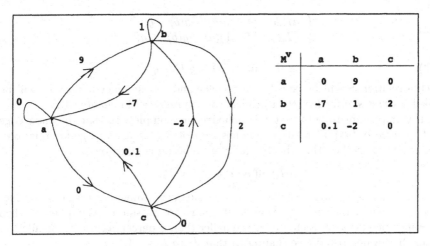

Figure 1.2: Example of valued relation

1.2 Introduction to the concept of semiorder

1.2.1 Definition

In many fields connected with decision-aid (economy, operations research, actuarial sciences, finance), the potential decisions (projects, candidates, ...) are evaluated on quantitative criteria, so that comparing decisions is equivalent to comparing numbers. The classical model underlying all these fields is the following:

> *if A is the set of potential decisions and g the function*
> *which associates a value g(a) to every element a of A,*
> *then, decision a "is at least as good as" decision b (aSb)*
> *iff g(a) ≥ g(b).*

Making the distinction between the relation "is strictly better than" and the relation "is as good as" (i.e. the asymmetric and symmetric parts of relation S) one obtains, $\forall a, b \in A$,

$$(1.1) \qquad \begin{cases} aP_sb & \text{iff} & g(a) > g(b), \\ aI_sb & \text{iff} & g(a) = g(b). \end{cases}$$

However, reflection suggests that it is not very reasonable to consider that a decision a is strictly better than b as soon as the value of a is higher than the value of b; the unavoidable imprecisions on the evaluations of the decisions often force to consider as equal, values which are very close to each other. This leads to the introduction of a positive threshold q (indifference, sensitivity or tolerance threshold) such that, $\forall a, b \in A$,

(1.2)
$$\begin{cases} aP_sb & \text{iff} \quad g(a) > g(b) + q, \\ aI_sb & \text{iff} \quad |g(a) - g(b)| \leq q, \end{cases}$$

or

(1.3)
$$aSb \quad \text{iff} \quad g(a) \geq g(b) - q.$$

Such a relation S will be called a *semiorder* and its asymmetric part P_s will be called a *strict semiorder* (their properties are presented in chapter 3).

It is important to note that S is reflexive and complete so that the knowledge of P_s ("strictly better" relation) implies that of S ("at least as good" relation); this is due to the fact that the "as good as" relation is given by

$$aI_sb \text{ iff } a\neg P_sb \text{ and } b\neg P_sa.$$

More generally, the threshold q may vary along the numerical scale of the values of g. This will be the case, for example, if a and b are considered as indifferent when the difference between their values is smaller than a percentage of the smallest of them. It can be proved (see chapter 3) that if, $\forall a, b \in A$,

(1.4)
$$g(a) > g(b) \quad \Rightarrow \quad g(a) + q_a \geq g(b) + q_b,$$

(where q_a and q_b are thresholds respectively associated to $g(a)$ and $g(b)$), then relation S is still a semiorder (which means that S has the same mathematical properties as when the threshold is constant).

The concept of semiorder is also encountered when the evaluation of each decision is an interval between a minimal (*pessimistic*) value and a maximal (*optimistic*) value. A possible attitude consists, in such a situation, to declare that *decision a is strictly better than b* if the interval associated to a lies entirely to the right of the interval associated to b. Both decisions are then considered as *indifferent* when their intervals have a non-empty intersection. If no interval is strictly included in another, we obtain again a semiorder; indeed, denoting $g(a)$, the left end point of the interval associated to a and $g(a) + q_a$ its right end point leads to the preceding situation (condition (1.4) being satisfied).

This introduction of the concept of semiorder through the comparison of intervals allows to give an idea of the two main properties of this structure (they will be more formally presented in chapter 3). These two properties are illustrated in figures 1.3 and 1.4

We will see that these properties are not only necessary but also sufficient to ensure that S is a semiorder (and P_s a strict semiorder), i.e. to ensure that S can be represented as in (1.3).

In summary, the following properties are equivalent (rigorous proofs can be found in chapter 3):

- S is a semiorder on A;

- P_s (asymmetric part of S) is a strict semiorder on A;

- there are a real-valued function g, defined on A, and a positive constant q such that, $\forall\ a, b \in A$,

(1.5)
$$aSb \quad \Longleftrightarrow \quad g(a) \geq g(b) - q;$$

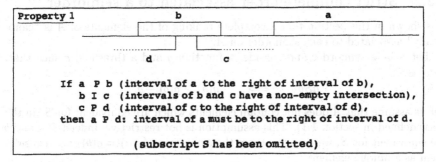

Figure 1.3: Property 1

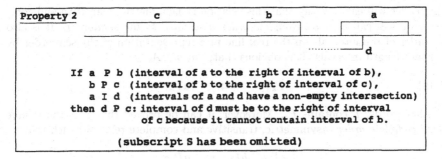

Figure 1.4: Property 2

- there are a real-valued function g, defined on A, and a positive constant q such that, $\forall\, a, b \in A$,

(1.6)
$$\begin{cases} aP_s b & \Longleftrightarrow & g(a) > g(b) + q, \\ aI_s b & \Longleftrightarrow & |g(a) - g(b)| \le q; \end{cases}$$

- there are two real-valued functions g' and q', defined on A, such that, $\forall\, a, b \in A$,

(1.7)
$$\begin{cases} aP_s b & \Longleftrightarrow & g'(a) > g'(b) + q'(b), \\ aI_s b & \Longleftrightarrow & \begin{cases} g'(a) \le g'(b) + q'(b), \\ g'(b) \le g'(a) + q'(a), \end{cases} \\ g'(a) > g'(b) & \Rightarrow & g'(a) + q'(a) \ge g'(b) + q'(b); \end{cases}$$

- the asymmetric part P_s and the symmetric part I_s of the reflexive relation S satisfy properties 1 and 2.

1.2.2   Strict complete order associated to a semiorder

It is shown in this section that a complete ranking of the elements of A is "canonically" associated to each semiorder on A.

Let S be a semiorder and consider a function g and a threshold q such that, $\forall\, a, b \in A$,

$$(1.8) \qquad aSb \quad \Longleftrightarrow \quad g(a) \geq g(b) - q.$$

Let us assume that A does not contain elements which are equivalent for S (in the sense defined in section 1.1). This assumption is not restrictive; indeed, if a and b are equivalent for S, we can give them the same value ($g(a) = g(b)$) and consider them as a unique element.

Let T be the relation defined by

$$(1.9) \qquad aTb \quad \Longleftrightarrow \quad g(a) > g(b).$$

The reader will easily verify that this relation is asymmetric, transitive and complete; we will call it *strict complete order associated to the semiorder*. It is also the order of the intervals on the real line in a representation of the semiorder by constant length intervals. It is obvious that, $\forall\, a, b \in A$,

$$(1.10) \qquad aP_sb \quad \Longrightarrow \quad aTb;$$

in other words, T contains P_s. Moreover, it can be proved that T is the unique *strict complete order* (asymmetric, transitive and complete relation) such that

$$(1.11) \qquad \begin{cases} aP_sb\,,\ bTc \ \Rightarrow\ aP_sc, \\ aTb\,,\ bP_sc \ \Rightarrow\ aP_sc; \end{cases}$$

it can also be proved that

$$(1.12) \qquad \begin{cases} aTb \text{ iff there is an element } c \text{ such that } aP_sc \text{ and } cI_sb \\ \text{or } aI_sc \text{ and } cP_sb. \end{cases}$$

Finally, given an asymmetric relation P, it will be a strict semiorder iff there is a strict complete order T such that

$$(1.13) \qquad \begin{cases} aPb\,,\ bTc \ \Rightarrow\ aPc, \\ aTb\,,\ bPc \ \Rightarrow\ aPc. \end{cases}$$

1.2.3   Matrix associated to a semiorder

We saw in the previous section that it was possible to univocally associate a strict complete order T to a semiorder S. Let us consider the matrix of relation P_s obtained by using T to associate the lines and columns of the matrix to the elements of A. Properties (1.11) mean that if an element of the matrix is equal to 1, then all the elements of the matrix which are to the right of and above this element must be equal to 1. We obtain a step-type matrix as illustrated in the next section. All 1's are obviously above the diagonal because the diagonal elements are null (P_s is asymmetric). With the same order T, the matrix associated to the semiorder S, will also be step-type, but with all the diagonal elements equal to 1 (S is reflexive).

Reciprocally, every step-type matrix with diagonal elements equal to 0 (resp. equal to 1) defines a strict semiorder (resp. a semiorder).

1.2.4 Example

Figure 1.5 is an example of semiorder on the set $A = \{a, b, c, d, e, f\}$ where the elements of A are represented by intervals of the real line; in figure 1.6, the same semiorder is represented by associated graphs and matrices.

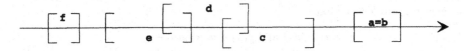

Figure 1.5: Interval representation of a semiorder

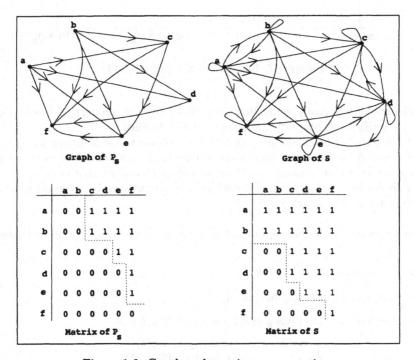

Figure 1.6: Graph and matrix representations

- Taking a threshold equal to 1, a numerical representation of the semiorder given in figure 1.5 is obtained by defining

$$g(f) = 0, \ g(e) = 2, \ g(d) = 3, \ g(c) = 4, \ g(b) = g(a) = 6;$$

- a and b are equivalent;
- for instance, we have cTd because cP_se and eI_sd.

1.3   Introduction to the concept of interval order

1.3.1   Definition

Consider the threshold model which was used to introduce the concept of semiorder but let us take the case where the threshold is variable and there are at least two elements a and b such that

(1.14) $$\qquad\qquad g(a) > g(b) \text{ and } g(a) + q_a < g(b) + q_b.$$

This situation corresponds to the case where, comparing intervals as in 1.2.1, one observes an interval which is completely included in another as in figure 1.7

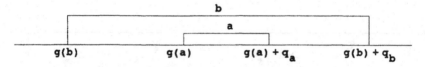

Figure 1.7: Illustration of relation (1.14)

     The reader will easily verify that, in such a situation, property 1 (see section 1.2.1) is verified but property 2 is not necessarily fulfilled.

     We will see (chapter 3) that property 1 is necessary and sufficient to be able to represent S by a model with a variable threshold; relation S is then called an *interval order*; as for semiorders, the knowledge of P_s implies that of S.

     A semiorder is an interval order which is representable by intervals none of which is included in another.

     In summary, the following properties are equivalent (formal proof in chapter 3):

- S is an interval order on A;

- P_s is a strict interval order on A;

- there are two functions g and q such that, $\forall \, a, b \in A$,

(1.15) $$\left\{ \begin{array}{l} aP_sb \iff g(a) > g(b) + q(b), \\ aI_sb \iff \left\{ \begin{array}{l} g(a) \leq g(b) + q(b), \\ g(b) \leq g(a) + q(a); \end{array} \right. \end{array} \right.$$

- the asymmetric part P_s and the symmetric part I_s of the reflexive relation S satisfy property 1 of section 1.2.1.

Note also that every semiorder is an interval order but that the reciprocal property is not true (example in the next section).

We will see in chapter 3 that two particular complete orders can be associated to an interval order, allowing to obtain a particular matrix representation but we will not present this property in this introductive chapter.

1.3.2 Example

An example of interval order is provided in figures 1.8 and 1.9.

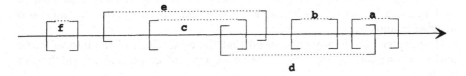

Figure 1.8: Interval representation

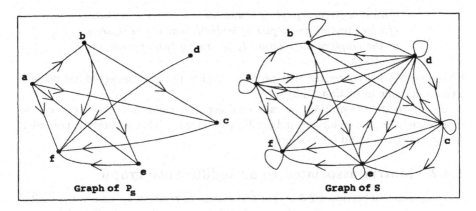

Figure 1.9: Graph representation

- A numerical representation is obtained by taking the following values for g and q

	a	b	c	d	e	f
g	9	7	3	4	2	0
q	2	1	2	6	4	1

- It is not a semiorder because some intervals are included in some others; as an exercise, the reader will verify that it is not possible to obtain the same graph of P_s with intervals having the property that no one is included in another (because property 2 is not satisfied).

1.4 Interval graph and indifference graph

1.4.1 Definition

An *interval graph* is a non-oriented graph with the property that every node can be represented by an interval of the real line in such a way that two nodes of the graph are connected by an edge if and only if the associated intervals have a non-empty intersection.

An *indifference graph* is an interval graph where the intervals associated to the nodes can be put on the line with the additional property that no interval is included in another.

The reader has already understood that the relations I_s of the interval order and of the semiorder respectively define an interval graph and an indifference graph.

Reciprocally, it can be shown that an interval graph (resp. indifference graph) can always be seen as the graph of the relation I_s associated to an interval order (resp. semiorder). This interval order (semiorder) is not unique. However it is interesting to note the following property:

> *if an indifference graph is connected (i.e. there is a chain of edges linking every pair of nodes), then it can be seen as the graph of the relation I_s of exactly two opposite semiorders.*

If the graph is not connected that property applies in each connected component (see an example in section 1.4.3).

So, up to the orientation, we may consider that there is a one-to-one correspondence between semiorders and indifference graphs. This property is not valid for interval graphs and interval orders.

1.4.2 Matrix associated to an indifference graph

A consequence of section 1.2.3 is that the matrix of relation I_s, associated to the semiorder S, obtained by using the associated strict complete order T to define the line and columns of the matrix, is such that all the 1's are symmetrically disposed along the diagonal and separated from the 0's by a step-type frontier, as illustrated in the next example (figure 1.10).

1.4.3 Example

Considering the connected component $\{c, d, e\}$, it is the indifference graph of exactly two opposite semiorders whose asymmetric parts are represented in figure 1.11.

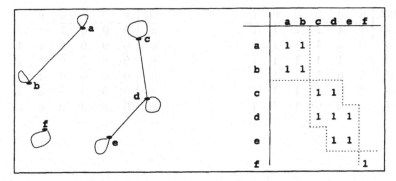

Figure 1.10: Indifference graph associated to the semiorder of figure 1.5

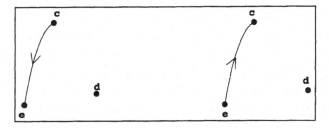

Figure 1.11: Two semiorders associated to the same indifference graph

1.5 Valued semiorder

Consider the valued relation defined by the following matrix:

	a	b	c	d	e	f
a	0.4	0.5	0.5	0.7	0.8	1
b	0.1	0.4	0.4	0.5	0.7	0.9
c	0.1	0.3	0.4	0.5	0.7	0.9
d	0	0.1	0.2	0.4	0.6	0.8
e	0	0.1	0.2	0.4	0.4	0.6
f	0	0	0.1	0.3	0.3	0.4

A peculiarity of the above matrix is that the values of its elements are increasing from left to right in each line and from bottom to top in each column. We will say that such a matrix is "step-type". Choosing a value α between 0 and 1 and changing to 1 every element $\geq \alpha$ and to 0 every element $< \alpha$ yields the matrix of a semiorder or of a strict semiorder, for every value α. Such a valued relation will be called a *valued semiorder*.

We give below the semiorder and the strict semiorder respectively obtained for $\alpha = 0.2$ and $\alpha = 0.6$.

0.2	a	b	c	d	e	f
a	1	1	1	1	1	1
b	0	1	1	1	1	1
c	0	1	1	1	1	1
d	0	0	1	1	1	1
e	0	0	1	1	1	1
f	0	0	0	1	1	1

0.6	a	b	c	d	e	f
a	0	0	0	1	1	1
b	0	0	0	0	1	1
c	0	0	0	0	1	1
d	0	0	0	0	1	1
e	0	0	0	0	0	1
f	0	0	0	0	0	0

HISTORICAL REVIEW AND APPLICATIONS

2.1 Some historical aspects

Let t_i be a cup of tea containing i milligrams of sugar. Any ordinary human being, comparing cups of tea, will generally consider that there is no difference between t_i and t_{i+1} (few people are able to perceive a difference of 1 milligram of sugar), and this, for every i. We say that a person is *indifferent* between t_i and t_{i+1}. However, she may have a preference for t_N over t_o (or the contrary) when N is large enough. This example, which shows that the indifference relation of an individual is not necessarily transitive, was introduced by Luce (Luce 1956). Before him, Armstrong 1939 and Georgescu–Roegen 1936, had already mentioned this phenomenon.

In the last century, the same phenomenon was pointed out by the psychologist Fechner (Fechner 1860): "*the discrimination relation between stimulus is generally not transitive: this can be explained by the concept of differential threshold* ".

This intransitivity (which is well represented by the notions of semiorder and indifference graph) is not typical of the psychological sciences. The situation is exactly the same when a physical characteristic is measured with an instrument; the following quotation from Poincaré 1905 is our translation of the French text in appendix.

Sometimes we are able to make the distinction between two sensations, while we cannot distinguish them from a third sensation. For example, we can easily make the distinction between a weight of 12 grams and a weight of 10 grams, but we are not able to distinguish each of them from a weight of 11 grams. This fact can symbolically be written

$$A = B, \qquad B = C, \qquad A < C.$$

This could be considered as a characterization of the physical continuum, as given by observation and experiments; this "contradiction" has been solved by the introduction of the mathematical continuum. The latter is a scale with an infinite number of levels, which do not overlap each other, as do the elements of the physical continuum.

The physical continuum is like a nebula whose elements cannot be perceived, even with the most sophisticated instruments; of course, with a good balance (instead of human sensation), it would be possible to distinguish 11 grams from 10 and 12 grams, so that we could write

$$A < B, \qquad B < C, \qquad A < C.$$

But one could always find other elements D and E such that

$$A = D, \qquad D = B, \qquad A < B,$$

$$B = E, \qquad E = C, \qquad B < C,$$

and the difficulty would be the same; only the mind can resolve it and the answer is the mathematical continuum.

Halphen 1955, studying the foundation of statistics, introduced the concept of *comparable set* and a notion of *equivalence relation* which was not necessarily transitive. For the reader's convenience, here is a translation of the original French text which can be found in appendix.

The experiments of Fechner have shown that, given three successive weights α, β, γ, it may happen that one can distinguish α from γ but can neither distinguish α from β nor β from γ. There exists a threshold under which a human being is not able to make any distinction between physical weights. This means that in the set of sensations $\{A, B, C\}$ given by the physical weights $\{\alpha, \beta, \gamma\}$ there exists

$$\left\{ \begin{array}{l} \text{an equivalence between } A \text{ and } B, \\ \text{an equivalence between } B \text{ and } C, \\ \text{a distinction between } A \text{ and } C. \end{array} \right.$$

In other words, the equivalence relation, in the set of psychological realities, is not necessarily transitive. It must be noted that this phenomenon has nothing to do with the "identity principle": the use of this principle would mean that we project, in the field of psychology, a scheme which was built before observing the psychological reality as it is. For example, if we accept that sensations can be measured by numbers and that equivalent sensations are measured by equal numbers, it follows from transitivity of equality in arithmetic that the psychological equivalence must also be transitive: but there is the question.

In fact, here is the only justifiable consequence which can be deduced from the transitivity of equality in arithmetic: as this transitivity is considered as essential for mathematical sciences, we can establish between the universe and a mathematical representation, a biunivocal correspondence where equivalence is represented by equality, only if equivalence is transitive in the universe. If the universe contains intransitive equivalences, then its mathematical representation (with the previous conditions) is simply impossible. Such a universe cannot be reduced to a quantitative representation. The experimental law of Fechner seems to show that this is the case for the real universe, because the psychological facts belong to that universe and have an action on it (as illustrated by the nuclear bomb).

If the universe contains "qualitative" realities, which are not reducible to quantities, that does not mean that a scientific and positive study is not possible. One even can use, for this study a new type of mathematics, "mathematics of quality"; if we adapt our system of mathematical symbols to the universe, then we will be able to say that the universe is reducible to mathematics; it is only a question of language. We will outline such an algebra of quality which constitutes, in our opinion, necessary preliminaries to mathematical statistics.

Let us consider a set of objects A, A', C, $\ldots$, with a certain operation, called equivalence and denoted $\asymp$ (the assertion $A \asymp A'$ will be considered as identical to $A' \asymp A$).

If equivalence is transitive, we can associate to this set of objects an "isomorphic" set $X_1, X_2, \ldots$ such that to each A corresponds one X, to each pair of non equivalent A's correspond two distinct X's and to each pair of equivalent A's corresponds one X.

If equivalence is not transitive, isomorphic reduction is not possible; we are so used to postulate transitivity of equivalence that we can hardly accept the contrary. It may happen that equivalence is transitive for only some A's; we assume that the isomorphic reduction has been realized for those elements. Let us assume that another relation, called "prevalence" and denoted $\succ$, is defined on the set, with the following properties:

$$\left\{ \begin{array}{l} \succ \text{ and } \asymp \text{ are exclusive,} \\ \succ \text{ is transitive.} \end{array} \right.$$

We don't assume that

$$A \asymp B, \qquad B \succ C \qquad \Rightarrow \qquad A \succ C$$

because this would exclude the possibility of having

$$B \asymp A, \qquad A \asymp C \qquad B \succ C,$$

and $\asymp$ would be transitive (which is not the case).

We say that a set E is "comparable" if, for every pair A, B of elements of E, we have

$$A \succ B \qquad or \qquad B \succ A \qquad or \qquad A \asymp B.$$

The elements of a comparable set are called "values". We must be very prudent with this name; it does not belong to the elements themselves, but to the association of these elements with the relations $\succ$ and $\asymp$;

*if $\asymp$ is not transitive in E, it is not possible to associate to each ele-
ment of E a symbol which would be called "its value", so that to two
equivalent elements would correspond "the same value". That is why
we really hesitate to use this very dangerous name of "values": they
are "improper values". If equivalence is transitive in E, then the term
"value" is natural; we then speak of "true values" and the set E is said
to be "ordered".*

*Let us now examine the important particular case of a comparable set
E which is not ordered (equivalence is not transitive) but which satifies
the two following conditions:*

1. $A' \asymp A,\ A \succ B \succ C \Rightarrow A' \succ C$.

*2. Given two equivalent elements $A' \asymp A$, it is possible to find ele-
ments X which are equivalent to A' and prevalent to A, or which
are equivalent to A and prevalent to A', but it is not possible to
find both types of elements simultaneously.*

*In this case, it is possible to order A and A' by reference to the X's,
and this order is transitive. The set E, which is not naturally ordered,
can be artificially ordered. To each element A, we can associate the
"interval" of its equivalents. If an addition operation has been defined
on the set E, this set is approximately measurable, i.e. it can be represented by points on an axis, with a threshold within which are lying
the points of the interval associated to A. It is for example the case for
Fechner's law.*

As can be seen, the concept of semiorder was strongly present in these "philosophical texts", and we encourage the interested people to read again Armstrong
1939, Fechner 1860, Georgescu–Roegen 1936, Goodman 1951, Guilbaud 1978,
Halphen 1955, Luce 1956, Poincaré 1905, Wiener 1919–1920 (about Wiener, see
also Fishburn and Monjardet 1992).

In the next paragraphs, we introduce some applications showing the great
variety of fields where semiorders and indifference graphs are encountered.

2.2 Semiorders in genetics

A chromosome is generally considered, by the geneticians, as a linear arrangement
of genes. To build the map of a chromosome (that is the linear succession of
the constituting genes), one knows that the nearer two genes, the less often they
recombine in case of separation.

Let A and B be two chromosomes; in their descendants, one will find, besides
chromosomes which are identical to A and B, new chromosomes C and D, con-
stituted from the genes of A and B as for instance in figure 2.1; C and D in turn
will generate new chromosomes.

Observing a large number of descendants will provide information on the frequency of combination of the genes, hence on the distance between them in the

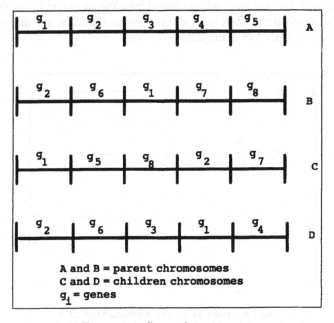

Figure 2.1: Some chromosomes

parent chromosomes. One obtains, for each chromosome, a matrix (with a number of lines and columns equal to the number of genes of the chromosome) whose element (i, j) is an increasing function of the distance between the corresponding genes.

The remaining problem is then to map the genes on a line in a manner which is compatible with the information contained in the matrix. This operation is possible only if the matrix represents a valued semiorder whose underlying complete order determines the linear succession of the genes.

In 1959, Benzer (Benzer 1962) asked himself whether it was possible, in a similar way, to study the fine structure of the gene and to determine whether this gene can also be considered as a linear arrangement of smaller elements. Unfortunately, the technique which was used for the chromosomes (interbreeding animals—flies in general—which are genetically rather similar and observing the characteristics of the descendants) could not be applied for the genes. Benzer then proposed to work on bacteria (this allows to treat very large populations and to very rapidly obtain successive generations) in the following manner. A population of genes is created from an initial gene; most of the elements of this population are identical to the "father" except for some of them which are called "mutants". A mutant is identical to the father except in a part which has been altered (the genetical information was not correctly copied). Biologists are able to recognize a mutant without knowing which part of the father has been modified.

The interesting idea is that two mutants issued from the same father can recompose
the father if and only if the alterations of these two mutants concern disjoined parts
of the gene, as illustrated in figure 2.2.

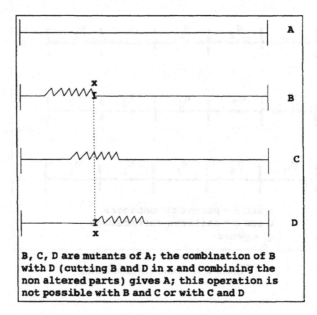

B, C, D are mutants of A; the combination of B
with D (cutting B and D in x and combining the
non altered parts) gives A; this operation is
not possible with B and C or with C and D

Figure 2.2: A gene and some mutants

As a consequence, from the paired combination of a large set of mutants having
the same father, one obtains information on the intersections between the different
parts of the father.

The question is thus to represent the parts of the gene by intervals on a line
in such a way to respect the previous information. This will be possible if and
only if the intersection graph of the parts of the gene is an interval graph (the
nodes are the parts and two nodes are connected by an edge iff the corresponding
parts intersect). Depending on the technique used, the parts of the gene may have
variable or constant lengths. In the latter case, the intersection graph of the parts
of the gene must be an indifference graph and we are facing the problem of finding
a semiorder compatible with the information obtained about the intersections of
the parts of the gene. In practice, this information is never perfect. If all the errors
consist in declaring as disjoined parts which, in reality, have an intersection, then
the problem to solve is the following: how to add a minimum number of edges
to a graph in order to obtain an indifference graph (or an interval graph). If all
the errors consist in declaring intersections which do not exist, the problem is to
delete a minimum number of edges from a graph in order to obtain an indifference
graph (or an interval graph). If errors can be of both types, there are usually
sure overlaps and sure non-overlaps, the rest being unknown; finding out whether

there exists an indifference graph (or an interval graph) whose edge set contains all sure overlaps and none of sure non-overlaps is called the *interval sandwich problem* after Golumbic and Shamir 1993.

Finally, it is interesting to observe that the map of the gene obtained in such a manner is not necessarily complete: certain alterations, concerning certain parts of the gene, may have been absent from the population of mutants. The concept of minimal representation of a semiorder (see chapter 4) could be very useful to detect potentially absent parts.

Research on such questions has remained quite active since they are important in the current programme aiming at mapping the human genome (Carrano 1988, Karp 1993, Nagaraja 1992). Recently Kaplan and Shamir 1996 investigated the above problems taking advantage of the fact that in the genetic context, the largest number of mutually overlapping parts is typically between 5 and 15. They have shown among other results that there exists a polynomial algorithm for the "interval sandwich problem" (with constant length intervals) provided the maximal number of mutually overlapping parts (i.e. the clique size of the graph) is bounded. They also showed that the problem of completing the graph with minimal clique size is equivalent to the well-known "bandwidth problem" which we present in the next subsection.

2.3 Semiorders, bandwidth and other difficult combinatorial problems

The *bandwidth problem* is the fourtieth in the famous list of NP-complete problems in graph theory established by Garey and Johnson 1979. Given a non-directed graph $G = (V, E)$ and a positive integer $K \leq |V| = n$, it consists in determining whether or not there is a linear arrangement of the set of nodes V with bandwidth K or less, i.e., a one-to-one mapping $f : V \longrightarrow \{1, \ldots, n\}$ such that, for each edge $\{u, v\} \in E$, $|f(u) - f(v)| \leq K$; in the related optimization problem, one has to find the minimal value of K for which the answer is positive.

In the genetic framework described in the previous section, the nodes could represent the parts of the gene and two nodes are linked by an edge whenever they are known to overlap. The problem is to determine whether there is a one-to-one representation of the n gene parts by the set of integer intervals of length K with left endpoints at $1, 2, \ldots, n$ which is compatible with the overlapping relation of the gene parts; in other terms one is looking for an interval representation such that the set of edges of the overlap graph is a subset of the intersection relation on the intervals.

The bandwidth problem is an important issue in other contexts such as matrix algebra (see e.g. Cuthill and McKee 1969); minimizing the bandwidth of a symmetric matrix amounts to minimizing by simultaneous row and column permutations, the number of "parallels" to the principal matrix diagonal in which there are non-zero elements.

Note that other problems studied in the genetic context are also well-known (and difficult). For instance, the minimal completion of a graph $G = (V, E)$ into an interval graph or an indifference graph $G' = (V, E')$ with $E' \supseteq E$, is also NP-hard (Garey and Johnson 1979, problem GT35). It corresponds in the genetic language, to finding a linear ordering of the gene pieces which is compatible with the known overlaps and minimizes the number of missing ones.

2.4   Semiorders in information storage

Let F be a file containing a set of elementary informations $a_1, a_2, \ldots, a_n$. Let Q be a set of "queries"; each query q_j requires a subset $q_j(F)$ of information contained in F. If the file has to be stored on a linear support (track of disk, book, tape record,...), one has advantage to group together all the information corresponding to each given query. The ideal situation is obtained when each subset $q_j(F)$ is stored on an interval of the linear support, as illustrated in the example of figure 2.3.

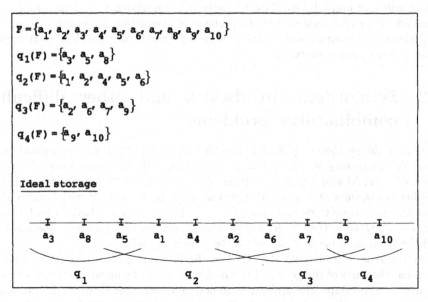

Figure 2.3: An example of information storage

The advantage of such a storage is that it is sufficient to know the places of the first and the last information for each query, to easily retrieve any information necessary for this query.

So, in this application, the problem is the following: given a set F and a set Q of parts of F, is it possible to rank the elements of F on a line in such a way that each element of Q is represented by an interval of the line. The underlying structure is thus an interval order on F; it is a semiorder if no element of Q is strictly included in another. The reader is referred to Eswaran 1975 for more detail.

2.5 Semiorders in decision-aid

It is well known that every decision-aid activity has to take into account unavoidable phenomena of uncertainty, imprecision and inaccurate determination. As pointed out by B. Roy (Roy 1985), these phenomena have various causes. First of all, the model is a communication tool and so, it is always a compromise between two goals: realism and tractability. Building a model inevitably involves simplifications, omissions and distortions which must lead the user to a great prudence. A second cause is the fact that the alternatives to be compared are often projects which involve unpredictable aspects. A third point is the impossibility, in general, to have precise evaluations of the alternatives, due to the measurement techniques or to ambiguities in the definition of what has to be measured.

From these considerations it appears that there is often some lack of realism in representing alternatives by precise numbers. A possible solution is to work with intervals or to introduce thresholds. The concept of semiorder is thus a very natural tool of decision-aid and preference modelling.

It is R.D. Luce (Luce 1956), who explicitly pointed out the phenomenon of intransitive indifference (with the example of the cup of tea, presented in section 2.1) and who proposed the introduction of the threshold model, presented in section 1.2.

However, that model was absent from decision-aid during a long time. Utility theory and value functions theory, which are the main tools of economists and of a lot of specialists in decision-aid are still based, today, on preference relations which are weak orders (assuming implicitly that the indifference threshold is equal to zero). It would be interesting to develop a utility theory which integrates the semiorder structure. This research avenue leads to non trivial problems; how to define, for instance, basic concepts like tradeoffs or preferential independence in the presence of thresholds. These aspects are considered in chapter 6.

It is the French multicriteria decision-aid school which first introduced the concept of threshold in decision-aid methods, via the definition of semi-criteria (i.e. threshold models), in methods such as ELECTRE or PROMETHEE (Roy and Bouyssou 1993, Brans and Vincke 1985). The software developed from these methods are designed for the assessment of indifference thresholds which are taken into account in the procedures.

As an example, let us consider the so-called "generalized criteria" of the PROMETHEE method (Brans and Vincke 1985). For each criterion j, a number $F_j(a, b)$ is associated to each (oriented) pair (a, b) of actions. This number is equal to zero when $g_j(a) \leq g_j(b)$, i.e. when a is not better than b for criterion j. It increases with the difference $g_j(a) - g_j(b)$ and is bounded by 1. In order to estimate the $F_j(a, b)$'s, the decision-maker is offered a choice between six shapes of curves, for each criterion. According to the way his preference increases with the difference $g_j(a) - g_j(b)$, the decision-maker sets, for each criterion, the shape of F_j and the associated parameters. These parameters are in fact indifference and preference thresholds. In reality, it is not difficult to see that each "generalized criterion" is a valued semiorder (and even a linear valued semiorder, as presented in chapter 5). It is also the case for the criteria which are used in ELECTRE III and IV (Roy and Bouyssou 1993).

Another example of threshold model in decision-aid can be found in the MAC-BETH method (Bana e Costa and Vansnick 1994). For each pair of actions a, b such that a is preferred to b, the decision-maker is asked to formulate an absolute judgement of difference of attractiveness by assigning the pair $\{a, b\}$ to one of six semantic categories: negligible, weak, moderate, strong, very strong and extreme. The procedure then tries to determine simultaneously a real number $v(a)$, for each action a, and six numbers $s_0 = 0, s_1, s_2, s_3, s_4, s_5$ such that

$$s_{k-1} < v(a) - v(b) \leq s_k$$

when $\{a, b\}$ has been assigned to the kth category $(k = 1, 2, \ldots, 5)$ and

$$v(a) - v(b) > s_5$$

when $\{a, b\}$ has been assigned to the category "extreme". This problem has a solution iff the underlying structure is a valued semiorder with particular properties which are presented in chapter 5 (valued semiorder representable with constant thresholds).

2.6   Semiorders in scheduling

Let us consider a set of "jobs" $a, b, c, \ldots$ which have to be done using some specific tools. These tools must be rented from an external supplier, so that we want to minimize the idle time of each tool. The ideal situation consists in scheduling the jobs (which cannot be simultaneously realized) in such a way that each tool is used without any interruption.

Let $A(x)$ be the set of jobs which require tool x. This tool will be used in an optimal way if and only if the jobs of $A(x)$ can be linearly ordered without any other job inserted. Thus the ideal situation corresponds to the case where each set $A(x)$ can be represented by an interval on the time axis.

Denoting $A(x) I A(y)$ the fact that some jobs require tools x and y, the problem is to verify whether the graph of relation I is an interval graph (it will be an indifference graph if no set $A(x)$ is included in another set $A(y)$).

The same situation occurs if the pair (jobs, tools) is replaced by the pair (projects, engineers) where one wants to avoid to continuously hire and dismiss engineers, or by the pair (courses, students) where one wants to avoid too many idle periods for the students.

2.7   Semiorders in mathematical programming

Although considerable progress has been made in the resolution of large mathematical programming problems, many real life problems lead to formulations that greatly exceed the computational limits. A solution is to try to partition the overall problem into manageable subproblems, but this is not always easy. Some large-scale problems present a structure that makes the interaction among the variables very hard to separate. Fortunately, these problems often contain special structure of which advantage can be taken by by building specific algorithms.

In particular, dynamic models often lead to large-scale problems: decisions must be made at several instants, but decisions made in any time period have impact upon other time periods. This situation leads to mathematical programs with staircase or block triangular structures, as illustrated in figure 2.4 (Bradley, Hax and Magnanti 1977).

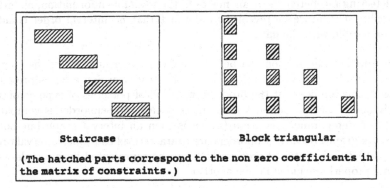

Staircase   **Block triangular**

(The hatched parts correspond to the non zero coefficients in the matrix of constraints.)

Figure 2.4: Staircase and block triangular structures in mathematical programming

The recognition of such patterns consists in finding a "good" numbering of the variables, that is the underlying order of a semiorder (where two variables are "indifferent" when they appear in a same constraint).

2.8 Interval orders and semiorders in archaeology

A major concern for archaeologists is to date events or historical periods. This can be done either in absolute manner, by locating them on the time scale, or by situating the periods relatively to each other. Archaeologists are particularly interested in the chronology of production periods for various types of artifacts (tools, potteries, fabrics,...). For example, Petrie 1899 studied the chronology of different types of potteries produced in prehistoric Egypt. Analysing the contents of 900 graves reveals that certain types of potteries were contemporaneous. If we suppose that each type of pottery has been produced during a certain time interval, finding potteries of two different types in the same grave means that the corresponding time intervals are not disjoined. The whole information about contemporaneity can hence be summarized in a non-oriented graph whose nodes represent the periods of production of each type of pottery and nodes are linked by an edge whenever potteries of the corresponding types have been found together in at least one grave. The chronology of the potteries production periods is an interval order compatible with the available elements of information, i.e. the just described non-oriented graph which is an interval graph or a partial interval graph provided there is no error or anomaly in the historical data.

Among the questions of interest to archaeologists are:

- finding all interval orders compatible with the contemporaneity relation encoded in the non-oriented graph and then trying to select the right chronology by using extraneous information or knowledge;

- detecting incoherencies or anomalies in the available information; anomalies could for instance be revealed by the fact that no interval order could be compatible with the data.

Some further information can be derived from closer analysis of the contemporaneity graph. If a configuration such as illustrated in figure 2.5 is present in the graph, this implies that the production period of potteries of type d contains properly one of the periods a, b or c. In that case, no semiorder is compatible with the contemporaneity relation (which is then an interval graph but not an indifference graph) because semiorders are characterized as proper interval orders (Roberts 1976), i.e. interval orders which can be represented by intervals which never are properly included in one another.

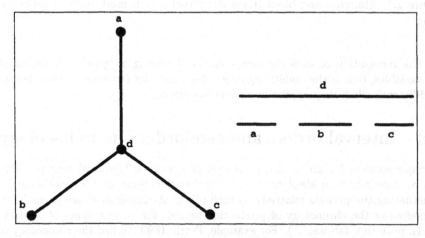

Figure 2.5: Forbidden configuration in the indifference relation of a semiorder

Note also that the *frequency* of presence of two types of potteries in the graves can be taken into account; the contemporaneity graph becomes a valued graph or valued binary relation. The corresponding information can be exploited in a manner that is outlined below in section 2.10.

Further detail on the topics above can be found in the book edited by Hodson, Kendall and Tautu 1971 and in the papers by Kendall (1963, 1969a, 1969b, 1971b, 1971a), Robinson 1951; see also Roberts 1976 and Roberts 1979.

2.9 Interval orders and semiorders in psychology

Besides the fundamental notion of *just-noticeable difference* (jnd), introduced by Fechner as a unit of difference of perception in sensory experiments and already mentioned in the beginning of the present chapter, there are further applications in psychology.

The problem of ordering time periods also occurs in this discipline, in particular in behavioral psychology. The psycho-affective development of the child can be described as a succession of periods characterized by particular behavioral traits. Psychologists have tried to elicit a natural order of succession of the different periods on the basis of the observation of a large number of children. Those in which coexist two or more behavior traits are recorded. The period associated with each behavior trait is represented by an interval and two intervals are disjoined whenever the two corresponding traits never appeared in the same observed individual. All interval orders whose indifference graph is the set of pairs of traits observed jointly in some child, are possible models for the chronology of the child development. (See Coombs and Smith 1973 for more detail).

2.10 Semiorders in seriation problems

One of the basic problems of data analysis to which applied researchers have paid attention for many years is the sequencing of objects along a continuum. A political scientist may wish to place legislators along a liberal-conservative axis, a psychologist may attempt to seriate subjects along a moral or developmental continuum, ... In the previous two sections, we have already considered that problem, but in a chronological context. Also the mapping of the genome dealt with above as our first application is a seriation problem.

In all of these cases, it is assumed that a set of objects can be sequenced in some linear fashion and the researcher's goal is to arrange these objects appropriately on the basis of the available information.

One of the most common types of data can be represented by a matrix (s_{ij}), where s_{ij} is some positive index of the proximity between objects i and j; the larger the index, the more similar the pair of objects. The basic problem is to construct a sequencing of the objects that is consistent with the ordinal part of the information contained in the proximity values. Let G_ϵ be the graph whose nodes represent the objects; i and j are linked by an edge iff $s_{ij} \geq \epsilon$. The aim of a seriation strategy is to produce a sequencing of the nodes which is "compatible" with the graph G_ϵ ("compatible" means here the following: if there is an edge between objects i and $i + k$, then there is also an edge between objects j and l, whenever j and l are included in the interval $[i, i + k]$). It turns out in fact that the existence of a compatible sequencing of the nodes is equivalent to G_ϵ being the symmetric part of a semiorder.

Intuitively, if the same sequencing of the nodes is compatible with G_ϵ, $\forall \epsilon$, then this sequencing is consistent with the ordinal information provided by (s_{ij}). This will be the case if the lines and columns of (s_{ij}) can be rearranged in order to obtain a so-called Robinson matrix (where the elements, in any row and column, never decrease as the entries are considered in sequence moving toward the main diagonal); considering the matrix $(K - s_{ij})$, where $K \geq s_{ij}$, $\forall\, i, j$, leads of course to a valued semiorder (the below-diagonal elements are set equal to zero without losing any information).

The desired sequencing can be produced, for example, by searching a maximal spanning tree; if s_{ij} can be transformed into a Robinson form, the maximum spanning tree consists of a single path and the sequence of nodes in the path defines the appropriate permutation of the rows and columns of s_{ij} needed to put the matrix into the proper form; more generally, the degree to which the maximum spanning tree does not form a path between two nodes is an indirect indication of non-seriability of the objects (see Hubert 1974 and Laporte 1987).

2.11 Semiorders in classification

Let us consider the symmetric *dissimilarity* matrix of figure 2.6; the greater the element (i, j) of that matrix, the less similar are objects i and j.

	a	b	c	d	e	f	g	h
a	.	2	2	1	12	12	12	12
b		.	2	7	12	12	12	12
c			.	7	12	12	12	12
d				.	12	12	12	12
e					.	3	3	2
f						.	3	10
g							.	10
h								.

Figure 2.6: Dissimilarity matrix

The problem is to group these elements in "homogeneous" classes (that is classes of similar elements). The hierarchical methods consist in first grouping the two most similar elements in a class; this class is then considered as a new element; the dissimilarity between this new element and the others is defined on the basis of the initial dissimilarities (there are several definitions which are commonly used in the applications). The two most similar elements are grouped again and the procedure ends up with all the elements in a single class. Such a method produces an *indexed hierarchy* yielding a family of reasonable partitions of the initial set in homogeneous classes.

We illustrate that procedure below on the example of figure 2.6, where the dissimilarity between an element and a class is defined as the "maximum" value of the dissimilarities between that element and the elements of the class. Afterwards, the same example is treated with "maximum" substituted by "minimum". Note that in the classical algorithm, only one new class is formed at each step by merging exactly two of the current classes. We use here a variant of this algorithm. Let δ denote the minimal value of the dissimilarity index between pairs of current classes; we look at all maximal subsets of current classes in which the degree of dissimilarity between all pairs of classes is equal to δ and for each such subset, we merge its components into a new class.

Hierarchical method with operator "maximum"

Level 1	ad	b	c	e	f	g	h
ad	.	7	7	12	12	12	12
b		.	2	12	12	12	12
c			.	12	12	12	12
e				.	3	3	2
f					.	3	10
g						.	10
h							.

Level 2	ad	bc	eh	f	g
ad	.	7	12	12	12
bc		.	12	12	12
eh			.	10	10
f				.	3
g					.

Level 3	ad	bc	eh	fg
ad	.	7	12	12
bc		.	12	12
eh			.	10
fg				.

Level 7	adbc	eh	fg
adbc	.	12	12
eh		.	10
fg			.

Level 10	adbc	ehfg
adbc	.	12
ehfg		.

Indexed hierarchy with operator "maximum"

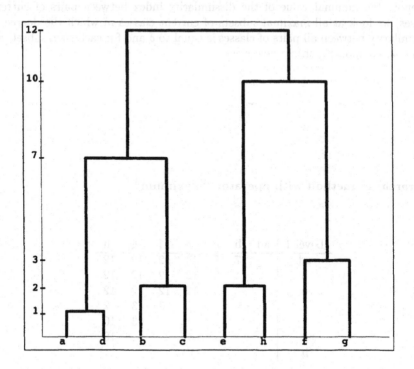

Hierarchical method with operator "minimum"

Level 1	ad	b	c	e	f	g	h
ad	.	2	2	12	12	12	12
b		.	2	12	12	12	12
c			.	12	12	12	12
e				.	3	3	2
f					.	3	10
g						.	10
h							.

Level 2	adbc	eh	f	g
adbc	.	12	12	12
eh		.	3	3
f			.	3
g				.

Level 3	adbc	ehfg
adbc	.	12
ehfg		.

Indexed hierarchy with operator "minimum"

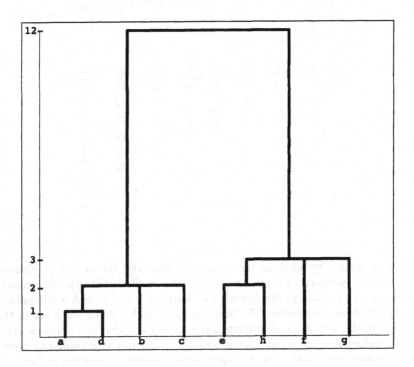

In fact, this kind of method is equivalent to changing some values of the initial matrix in such a way that disjoined classes can be obtained. It is interesting to reconstitute the dissimilarity matrices which, without any modification, would have led to the preceding hierarchies. These matrices are represented below (they are trivially obtained from the preceding indexed hierarchies). In each matrix, we have encircled the values which have changed relatively to the initial data.

Max	a	b	c	d	e	f	g	h
a	.	[7]	[7]	1	12	12	12	12
b		.	2	7	12	12	12	12
c			.	7	12	12	12	12
d				.	12	12	12	12
e					.	[10]	[10]	2
f						.	3	10
g							.	10
h								.

Min	a	b	c	d	e	f	g	h
a	.	2	2	1	12	12	12	12
b		.	2	[2]	12	12	12	12
c			.	[2]	12	12	12	12
d				.	12	12	12	12
e					.	3	3	2
f						.	3	[3]
g							.	[3]
h								.

These dissimilarity matrices have a special property; for any three elements x, y, z, if the dissimilarity between x and y is the smallest, then the dissimilarity between x and z is equal to the dissimilarity between y and z; such a dissimilarity index is called an *ultrametric*. Building an indexed hierarchy on a set is thus equivalent to the definition of an ultrametric on this set. One can of course try to find an ultrametric which is "as close as possible" to the initial dissimilarity matrix. The ultrametric provided by the operator maximum (resp. minimum) is the closest one among all the ultrametrics obtained by increasing (resp. decreasing) the values of the initial matrix. The ultrametric given below (with its indexed hierarchy) is the closest one for the least squares method.

Least squares	a	b	c	d	e	f	g	h
a	.	2	2	[7]	12	12	12	12
b		.	2	7	12	12	12	12
c			.	7	12	12	12	12
d				.	12	12	12	12
e					.	3	3	[10]
f						.	3	10
g							.	10
h								.

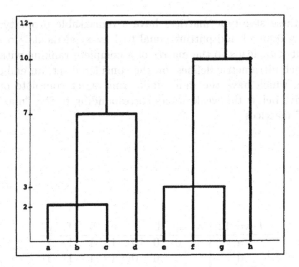

Remark that the ultrametric allows to rank the elements along an axis in such a way that, starting from a point, the farther an element is from that point, the greater the dissimilarity. As a consequence, writing the matrices of the dissimilarities with that ranking of the elements on the lines and columns leads to step-type matrices. This is shown in the above matrix for the least squares method and in the matrices below for the max and the min methods.

Max	a	d	b	c	e	h	f	g
a	.	1	7	7	12	12	12	12
d		.	7	7	12	12	12	12
b			.	2	12	12	12	12
c				.	12	12	12	12
e					.	2	10	10
h						.	10	10
f							.	3
g								.

Min	a	d	b	c	e	h	f	g
a	.	1	2	2	12	12	12	12
d		.	2	2	12	12	12	12
b			.	2	12	12	12	12
c				.	12	12	12	12
e					.	2	3	3
h						.	3	3
f							.	3
g								.

Moreover, these step-type matrices have a remarkable property: choosing a value α between 0 and 1 and putting equal to 1 every element $\geq \alpha$ and equal to 0 every element $< \alpha$, leads to the matrix of a complete ranking with ties ("weak order"). So, each ultrametric defines, on the considered set, an embedded family of weak orders which have the same underlying strict complete order. As an example, we give below the weak orders corresponding to the "max" and to the "least squares" matrices.

Max matrix

Level 12	a	d	b	c	e	h	f	g
a	.	0	0	0	1	1	1	1
d		.	0	0	1	1	1	1
b			.	0	1	1	1	1
c				.	1	1	1	1
e					.	0	0	0
h						.	0	0
f							.	0
g								.

Level 10	a	d	b	c	e	h	f	g
a	.	0	0	0	1	1	1	1
d		.	0	0	1	1	1	1
b			.	0	1	1	1	1
c				.	1	1	1	1
e					.	0	1	1
h						.	1	1
f							.	0
g								.

Level 7	a	d	b	c	e	h	f	g
a	.	0	1	1	1	1	1	1
d		.	1	1	1	1	1	1
b			.	0	1	1	1	1
c				.	1	1	1	1
e					.	0	1	1
h						.	1	1
f							.	0
g								.

Level 3	a	d	b	c	e	h	f	g
a	.	0	1	1	1	1	1	1
d		.	1	1	1	1	1	1
b			.	0	1	1	1	1
c				.	1	1	1	1
e					.	0	1	1
h						.	1	1
f							.	1
g								.

Level 2	a	d	b	c	e	h	f	g
a	.	0	1	1	1	1	1	1
d		.	1	1	1	1	1	1
b			.	1	1	1	1	1
c				.	1	1	1	1
e					.	1	1	1
h						.	1	1
f							.	1
g								.

Least squares matrix

Level 12	a	b	c	d	e	f	g	h
a	.	0	0	0	1	1	1	1
b		.	0	0	1	1	1	1
c			.	0	1	1	1	1
d				.	1	1	1	1
e					.	0	0	0
f						.	0	0
g							.	0
h								.

Level 10	a	b	c	d	e	f	g	h
a	.	0	0	0	1	1	1	1
b		.	0	0	1	1	1	1
c			.	0	1	1	1	1
d				.	1	1	1	1
e					.	0	0	1
f						.	0	1
g							.	1
h								.

Level 7	a	b	c	d	e	f	g	h
a	.	0	0	1	1	1	1	1
b		.	0	1	1	1	1	1
c			.	1	1	1	1	1
d				.	1	1	1	1
e					.	0	0	1
f						.	0	1
g							.	1
h								.

Level 3	a	b	c	d	e	f	g	h
a	.	0	0	1	1	1	1	1
b		.	0	1	1	1	1	1
c			.	1	1	1	1	1
d				.	1	1	1	1
e					.	1	1	1
f						.	1	1
g							.	1
h								.

In the literature on classification, valued step-type matrices are called *robinsonian dissimilarities* (Bertrand 1992). The interest for such dissimilarities has increased since the introduction of the so-called pyramidal representation, which generalizes indexed hierarchies in that they allow for non-disjoined classes. With this approach it is possible to visualize the data of a dissimilarity matrix (as done in an indexed hierarchy) in bringing much less modifications to the initial data. In our example (figure 2.6), reordering the elements leads to a robinsonian matrix, and, without any modification, to a pyramidal representation as shown below.

	d	a	b	c	h	e	f	g
d	.	1	7	7	12	12	12	12
a		.	2	2	12	12	12	12
b			.	2	12	12	12	12
c				.	12	12	12	12
h					.	2	10	10
e						.	3	3
f							.	3
g								.

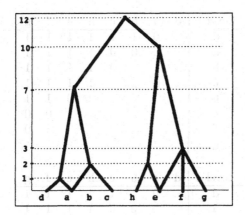

When a matrix is robinsonian, choosing α between 0 and 1 and putting equal to 1 every element $\geq \alpha$ and equal to 0 every element $< \alpha$, leads to the matrix of a strict semiorder. A robinsonian matrix is thus equivalent to an embedded family of strict semiorders which have the same underlying strict complete order. Hence the interest of the semiorders in classification.

Strict semiorders associated to the previous matrix

Level 12	d	a	b	c	h	e	f	g
d	.	0	0	0	1	1	1	1
a		.	0	0	1	1	1	1
b			.	0	1	1	1	1
c				.	1	1	1	1
h					.	0	0	0
e						.	0	0
f							.	0
g								.

Level 10	d	a	b	c	h	e	f	g
d	.	0	0	0	1	1	1	1
a		.	0	0	1	1	1	1
b			.	0	1	1	1	1
c				.	1	1	1	1
h					.	0	1	1
e						.	0	0
f							.	0
g								.

Level 7	d	a	b	c	h	e	f	g
d	.	0	1	1	1	1	1	1
a		.	0	0	1	1	1	1
b			.	0	1	1	1	1
c				.	1	1	1	1
h					.	0	1	1
e						.	0	0
f							.	0
g								.

Level 3	d	a	b	c	h	e	f	g
d	.	0	1	1	1	1	1	1
a		.	0	0	1	1	1	1
b			.	0	1	1	1	1
c				.	1	1	1	1
h					.	0	1	1
e						.	1	1
f							.	1
g								.

Note that, as in the previous section, we assumed here, without any loss of information, that the diagonal and below-diagonal elements are null.

2.12 Semiorders and rough sets

A rough set is basically a pair of a lower and an upper approximations for a set. Typically, rough sets theory provides tools for extracting a description of classes of individuals in terms of some of their characteristics (Pawlak 1991). In medical diagnosis, rough sets analysis can be used to extract rules based on clinical symptoms (and their degree of seriousness) that lead to the conclusion that a patient does or does not have a given disease. This is done on the basis of a learning set of patients described by a set of medical descriptors and suffering or not of the disease. The technique compares with discriminant analysis but applies in a different context, with non-numeric or imprecisely defined values of the attributes.

In the basic version, the domain of variation of each attribute is partitioned in a (small) number of classes of values which are considered indiscernible or equivalent. Two individuals belong to the same *elementary set* when their descriptors on each attribute have indiscernible values. Elementary sets are hence associated with specific classes of equivalent values on each attribute. The category of interest (e.g. the category of ill persons) is approximated by the union of all elementary sets it contains (lower approximation) and the union of all elementary sets it intersects (upper approximation). From there, algorithms are available to obtain minimal sets of rules corresponding respectively to the lower and the upper approximations (resp. "deterministic" and "non-deterministic" rules).

Two different extensions of the above approach have recently been proposed; they both consist to allow for *overlapping* classes of values on each attribute instead of using a *partition* in equivalence classes. Such overlapping classes naturally arise as indifference classes of semiorders. In the first extension (Słowiński 1992), due to the often unrealistic character of a sharp definition of indiscernibility classes, it is proposed to consider an inner and an outer indiscernibility class respectively included in and containing the original indiscernibility class; they are said to correspond respectively to strong and weak indiscernibility and are indeed the indifference classes of two semiorders.

In another extension (Słowiński and Vanderpooten 1995), the overlapping classes are defined on the basis of a (non-necessarily symmetric) *similarity* relation: x is similar to y w.r.t. an attribute if the absolute difference of their values is below a certain threshold (depending on y). If the threshold is constant, the similarity classes are again the indifference classes of a semiorder. Note that in this work, the construction of the elementary sets is viewed as a problem of *aggregation* of similarity relations (not necessarily in a conjunctive manner) which amounts, in the particular case of constant thresholds, to aggregating semiorders (or their symmetric parts).

2.13 Semiorders and fuzzy sets

Modelling imprecision has been a major motivation for the introduction of fuzzy sets (Zadeh 1965). Consider a set of objects which can be evaluated in an imprecise manner with respect to some point of view. One way of representing imprecision is by means of intervals and a calculus has been developed for intervals (Moore 1966; see also Hansen 1992, for a more recent reference). Fuzzy numbers and intervals generalize and enrich simple intervals as models for imprecision.

In the example of figure 2.7, two objects A and B are represented by means of trapezoidal fuzzy intervals. Eventually, the fuzzy interval associated with A represents the possible ending time of task A and the model can be interpreted as follows: task A will certainly terminate during the interval of time (t_1, t_4) but most likely between t_2 and t_3.

Comparing the ending times of the two tasks A, B, eventually represented in figure 2.7, one may be interested in the *possibility* for A to finish before B. The *possibilistic* interpretation was introduced by Zadeh 1978 and further developed by Dubois and Prade (see e.g. Dubois and Prade 1988; see also Klir and Bo Yuan 1995). The possibility for A to finish at least as late as B is defined by

$$\text{Pos}(A \geq B) = \sup\{\min\left[\mu_A(t), \mu_B(s)\right]; \text{ such that } t \geq s\},$$

with $\mu_A(t)$ (resp. $\mu_B(s)$), the degree of confidence in t (resp. s)as a final date for task A (resp. B). In the situtation illustrated in figure 2.7, $\text{Pos}(A \geq B)$ is just the height at which the righthand side of A meets the lefthand side of B, i.e. the second coordinate of point α (about .4). Conversely, the possibility for B to finish at least as late as A is 1. If only the ordering of the possibility values is considered meaningful, the whole information about the comparison of a set of

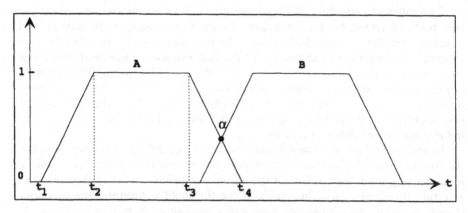

Figure 2.7: Two fuzzy intervals

objects $A, B, C, \ldots$, represented by fuzzy intervals can indeed be summarized by a chain of interval orders or semiorders. Namely, those obtained by cutting the fuzzy intervals by well-chosen horizontal lines. This is illustrated in figure 2.8.

The cuts are made at the heights where righthand sides and lefthand sides of fuzzy intervals cross each other (or slightly above those heights). If only the ordering of the possibility indices is meaningful, the three semiorders obtained by cutting the fuzzy intervals convey the same information as the full trapezoids. The only thing that really matters is the intersection relation of the intervals in each cut. Note that the intervals associated with the different cutting levels of a fuzzy interval are ordered by inclusion. In case the trapezoids all have equal upper bases and equal lower bases, all cuts yield semiorders.

Chains of interval orders and semiorders will be studied in chapter 5 together with valued interval orders and semiorders i.e. binary relations with a numerical value associated with each arc. In case the value associated to arc (A, B) is the possibility degree $\mathrm{Pos}(A \geq B)$, the cuts of the resulting valued relation are exactly the chain of interval orders (or semiorders) obtained by cutting the fuzzy intervals (see Roubens and Vincke 1988).

2.14   Semiorders and a theory of evolution of rationality

Falmagne and Doignon have recently proposed a new theory which accounts for the evolution of the preference of a subject on a set of objects (Falmagne and Doignon 1995). In the course of time, the subject receives elementary pieces of information from the outside world. These "tokens" of information each concern a specific pair of objects and perturb the current relation of preference by eventually adding a pair to or removing it from the current relation (not all perturbations actually produce a change; in the theory they are supposed to occur randomly).

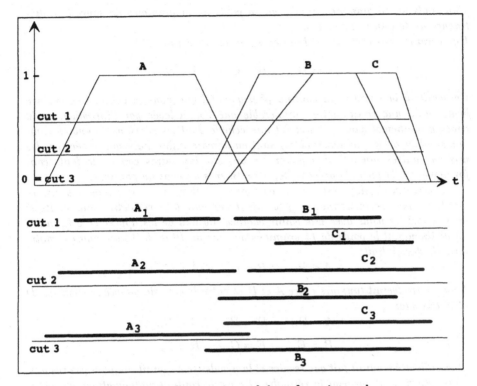

Figure 2.8: Comparaison of three fuzzy intervals

Considering the set of all semiorders on the set of objects as the space of potential preference structures, the authors conceive the preference evolution as a trajectory in that space. The cornerstone of the theory is the following interesting fact (established in Doignon and Falmagne 1994): any semiorder on a finite set of objects can be obtained from any other one through a sequence of elementary changes which

- transform a semiorder into a semiorder

- consist of removing a pair from or adding a pair to the previous semiorder.

2.15 Appendix

Text of Poincaré (Poincaré 1905)

"Il arrive que nous sommes capables de distinguer deux impressions l'une de l'autre, tandis que nous ne saurions distinguer chacune d'elles d'une même troisième. C'est ainsi que nous pouvons discerner facilement un poids de 12 grammes

d'un poids de 10 grammes, tandis qu'un poids de 11 grammes ne saurait se distinguer ni de l'un ni de l'autre.

Une pareille constatation, traduite en symboles s'écrirait:

$$A = B, \qquad B = C, \qquad A < C.$$

Ce serait là la formule du continu physique, tel que nous le donne l'expérience brute, d'où une contradiction intolérable que l'on a levée par l'introduction du continu mathématique. Celui-ci est une échelle dont les échelons (nombres commensurables ou incommensurables) sont en nombre infini mais sont extérieurs les uns aux autres, au lieu d'empiéter les uns sur les autres comme le font, conformément à la formule précédente, les éléments du continu physique.

Le continu physique est pour ainsi dire une nébuleuse non résolue, les instruments les plus perfectionnés ne pourraient parvenir à la résoudre; sans doute si on évaluait les poids avec une bonne balance, au lieu de les apprécier à la main, on distinguerait le poids de 11 grammes de ceux de 10 et de 12 grammes et notre formule deviendrait:

$$A < B, \qquad B < C, \qquad A < C.$$

Mais on trouverait toujours entre A et B et entre B et C de nouveaux éléments D et E tels que:

$$A = D, \qquad D = B, \qquad A < B,$$
$$B = E, \qquad E = C, \qquad B < C,$$

et la difficulté n'aurait fait que reculer et la nébuleuse ne serait toujours pas résolue; c'est l'esprit seul qui peut la résoudre et c'est le continu mathématique qui est la nébuleuse résolue en étoiles."

Text of Halphen (Halphen 1955)

"Il résulte des expériences classiques de Fechner que, étant donnés trois poids successifs α, β, γ, il arrive que l'on puisse distinguer α de γ sans pouvoir cependant distinguer α de β ni β de γ. Il existe entre les poids physiques un "seuil" au dessous duquel l'organisme d'un sujet donné ne sait pas faire de distinction. Evitant, comme il a été dit, de projeter les poids physiques α, β, γ, dans les sensations A, B, C, nous devons dire qu'en ce qui concerne ces sensations, il existe:

> *entre A et B, entre B et C, une relation d'équivalence,*
> *entre A et C, une relation de distinction.*

Il apparaît ainsi que la relation d'équivalence entre réalités psychologiques n'est pas nécessairement transitive.

Que l'on ne se récrie pas au nom du principe d'identité: nous avons suffisamment expliqué que la notion d'identité n'a rien à voir ici; si on l'invoque c'est qu'encore une fois on projette dans le domaine psychologique un schéma qu'on s'est fabriqué avant d'observer la réalité psychologique telle qu'elle est. Par exemple, on a admis d'avance que les sensations sont mesurables par des nombres, les sensations équivalentes étant mesurées par des nombres égaux; et comme l'égalité arithmétique est transitive on en conclut que l'équivalence psychologique doit l'être aussi: mais c'est précisément là toute la question.

En toute rigueur, voici la seule conséquence légitime que l'on puisse tirer du caractère transitif de l'égalité arithmétique: comme cette transitivité est regardée comme essentielle à la science mathématique, on ne peut établir entre l'univers et un schéma mathématique de correspondance biunivoque où l'équivalence soit représentée par une égalité, que si dans l'univers toute équivalence est transitive. Si l'univers contient en fait des équivalences non transitives, sa représentation par un schéma mathématique du type précédent est impossible. Autrement dit encore, un tel univers n'est pas totalement réductible à un schéma quantitatif. La loi expérimentale de Fechner semble bien démontrer que tel est le cas pour l'univers réel, car les faits psychologiques font partie de cet univers, et agissent sur lui (à une échelle que la bombe atomique a rendue appréciable).

Si l'univers contient des réalités "qualitatives" non réductibles à la quantité cela ne signifie pas qu'on doive renoncer à en faire une étude scientifique et même parfaitement positive. On pourra même, si l'on veut, utiliser pour cette étude une espèce de mathématique de type nouveau, une "mathématique de la qualité"; en commençant par dilater notre système de symboles "mathématiques" aux dimensions de l'univers, on pourra bien dire que l'univers est réductible aux mathématiques; ce n'est plus alors qu'affaire de langage.

Nous allons esquisser une telle algèbre de la qualité, prélude nécessaire, à notre avis, de la statistique mathématique.

Considérons un ensemble d'"êtres" A, A', C, ..., auxquels on a associé une certaine opération, nommée équivalence, qui se notera: $A \asymp A'$ (et qui est supposée identique à sa réciproque $A' \asymp A$).

Si l'équivalence est transitive, on peut faire correspondre à l'ensemble précédent un ensemble dit "isomorphe" $X_1, X_2, \ldots$ tel qu'à tout A corresponde un X, qu'à deux A non équivalents correspondent deux X distincts et qu'à deux A équivalents correspondent un seul X.

Mais si l'équivalence n'est pas transitive, cette réduction isomorphique n'est pas possible: c'est parce qu'on a l'habitude de la postuler qu'on se scandalise à la pensée d'une équivalence non transitive.

L'équivalence peut d'ailleurs être transitive pour une partie seulement des A: dans l'avenir nous supposerons d'ordinaire que pour ceux-là la réduction a été effectuée. Supposons maintenant qu'en outre de l'équivalence soit définie une certaine relation nommée "prévalence" $(A \succ B)$, possédant les propriétés suivantes: elle est exclusive de l'équivalence, ainsi que de sa réciproque (sa converse se nommera "postvalence" notée: $B \prec A$). Bien que ce ne soit probablement pas toujours le cas, nous supposerons la prévalence transitive:

$$\text{de } A \succ B \text{ et } B \succ C, \text{ on tirera: } A \succ C.$$

En général la prévalence ne commute pas avec l'équivalence, autrement dit, de $A \asymp B$ et $B \succ C$, on ne peut tirer $A \succ C$, car ce serait exclure la possibilité d'avoir simultanément: $B \asymp A$, $A \asymp C$, $B \succ C$; en ce cas l'équivalence serait transitive, ce qu'on ne suppose pas.

Si maintenant, étant donnés deux êtres quelconques d'un certain ensemble E, il existe toujours entre eux soit équivalence soit prévalence (ou postvalence), l'ensemble sera dit "comparable". Eu égard à ces relations d'équivalence et de prévalence, les éléments de l'ensemble seront alors nommés des "grandeurs".

Ce nom de grandeurs ne doit pas nous faire illusion: il n'appartient pas aux éléments eux-mêmes, mais à l'association de ces éléments avec leurs relations de comparaison; en outre, si l'équivalence n'est pas transitive dans l'ensemble nous savons qu'on ne peut associer à chaque élément un symbole qui constituerait "sa grandeur", de telle sorte qu'à deux éléments équivalents corresponde une "même grandeur". C'est donc avec beaucoup d'hésitations que nous acceptons d'employer ce nom dangereux de "grandeurs": il s'agit de grandeurs improprement dites. Si au contraire l'équivalence est transitive le terme de grandeur ne fait plus de difficulté: nous parlerons alors de grandeurs vraies, et l'ensemble E sera dit "ordonné".

Examinons un cas particulier important, celui d'un ensemble E comparable mais non ordonné (c'est-à-dire dans lequel l'équivalence n'est pas transitive) satisfaisant aux deux conditions suivantes:

1. *De $A' \asymp A$ et de $A \succ B \succ C$, on peut tirer $A' \succ C$.*

2. *Etant donné un couple quelconque d'éléments équivalents $A' \asymp A$, on peut trouver des X équivalents à A' et prévalents à A, ou bien on en peut trouver d'équivalents à A et prévalents à A', mais non les deux à la fois.*

Dans ces conditions, on peut ordonner A et A' par référence aux X, et cet ordre est transitif. L'ensemble E, dans ce cas, bien que non ordonné par nature, est ainsi ordonnable par un artifice. A chaque élément A on peut associer l'"intervalle" (ordonné comme il vient d'être dit) de ses équivalents. Si l'on a défini dans l'ensemble E une addition (comme nous l'expliquons ci-dessous), un tel ensemble est alors approximativement mesurable, c'est-à-dire qu'il est représentable par des points d'un axe avec un seuil E à l'intérieur duquel se trouvent les points figuratifs de l'intervalle associé à A. C'est ce qui se passe pour la loi de Fechner par exemple.

3

BASIC CONCEPTS AND
DEFINITIONS

We give in this chapter the rigorous definitions of the concepts we will use in the sequel of the book. Some definitions were already presented in chapter 1 but we have repeated them here in order to have all the basic definitions gathered in a same chapter and, also, because the purpose of chapter 1 was to give an intuitive introduction to the concepts while this chapter is the mathematical introduction of the book. The reader can also refer to Fishburn 1985, Monjardet 1978 or Roubens and Vincke 1985.

3.1 Binary relations

Let A denote a finite set of elements $a, b, c, \ldots$ and $|A|$, its number of elements. A *binary relation* S on the set A is a subset of the cartesian product $A \times A$, that is, a set of ordered pairs (a, b) such that a and b belong to A: $S \subset A \times A$. If the ordered pair (a, b) is in S, we denote $(a, b) \in S$ or aSb; if not, $(a, b) \notin S$ or $a \neg Sb$.

Let S and T be two relations on the same set A. The following notations will be used.

$$S \subset T \text{ iff } aSb \Longrightarrow aTb, \forall a, b \in A (\text{inclusion}),$$

$$a(S \cup T)b \text{ iff } aSb \text{ or (inclusive) } aTb (\text{union}),$$

$$a(S \cap T)b \text{ iff } aSb \text{ and } aTb (\text{intersection}),$$

$$aSTb \text{ iff } \exists c \in A : aSc \text{ and } cTb,$$

$$aS^2b \text{ iff } \exists c \in A : aSc \text{ and } cSb.$$

A binary relation S on the set A is

- *reflexive* iff $a \, S \, a, \forall a \in A$,
- *irreflexive* iff $a \, \neg S \, a, \forall a \in A$,
- *symmetric* iff $a \, S \, b \Longrightarrow b \, S \, a, \forall a, b \in A$,
- *antisymmetric* iff $a \, S \, b \Longrightarrow b \, \neg Sa, \forall a, b \in A$ such that $a \neq b$,

49

– *asymmetric* iff $a\,S\,b \implies b\,\neg S a,\ \forall a, b \in A$,

– *complete* iff $a\,S\,b$ or $b\,S\,a,\ \forall a, b \in A$ such that $a \neq b$,

– *transitive* iff $a\,S\,b, b\,S\,c \implies a\,S\,c,\ \forall a, b, c \in A$,

– *negatively transitive* iff $a\,\neg S b, b\,\neg S c \implies a\,\neg S c, \forall a, b, c \in A$,

– an *equivalence relation* iff it is reflexive, symmetric and transitive,

– a *strict partial order* iff it is asymmetric and transitive,

– a *partial order* iff it is reflexive, antisymmetric and transitive,

– a *partial preorder* or simply a *preorder* iff it is reflexive and transitive.

3.2 Asymmetric and symmetric parts of a relation

Given a binary relation S on a set A, we respectively denote P_S and I_S the asymmetric and the symmetric parts of S;

(3.1) $$\begin{cases} a\,P_S\,b & \iff a\,S\,b \text{ and } b\,\neg S\,a, \\ a\,I_S\,b & \iff a\,S\,b \text{ and } b\,S\,a. \end{cases}$$

It is clear that $S = P_S \cup I_S$. When no confusion is possible, P_S and I_S will be replaced by P and I.

3.3 Equivalence relation associated to a relation

Given a binary relation $S = P \cup I$, the relation E defined by

(3.2) $$a\,E b \text{ iff } \forall c \in A, \begin{cases} a\,S\,c & \iff b\,S\,c, \\ c\,S\,a & \iff c\,S\,b, \end{cases}$$

is clearly an equivalence relation.

3.4 Important conventions

If not otherwise mentioned, a relation denoted by I_S (or I) will always be considered as reflexive. Moreover, a reflexive relation will sometimes be denoted by a single letter (S for example) and sometimes by the ordered pair (P_S, I_S) or (P, I); in the latter case, P and I respectively represent the asymmetric and the symmetric (and reflexive) parts of the (reflexive) relation $P \cup I = S$.

We will assume that two different elements of A are never equivalent. This is not restrictive as one can always consider the quotient of A by E, i.e. the set of equivalence classes defined by E.

3.5 Complement, converse and dual of a relation

The complement S^c, the converse $\overline{S}$ and the dual S^d are respectively defined by

(3.3.)
$$\begin{cases} a\,S^c\,b & \text{iff} & a\,\neg S\,b, \\ a\,\overline{S}\,b & \text{iff} & b\,S\,a, \\ a\,S^d\,b & \text{iff} & b\,\neg S\,a. \end{cases}$$

Remark that

$$\begin{cases} P_S = S \cap S^d, \\ I_S = S \cap \overline{S}. \end{cases}$$

3.6 Graph and matrix representations of a binary relation

A relation S on a set A can be represented by a directed graph (A, S) where A is the set of nodes and S the set of arcs. When $a\,S\,a$, a loop is drawn at the node a. A *path* in a graph on A is an ordered sequence of arcs of the form $(a_0, a_1), (a_1, a_2), \ldots, (a_{k-1}, a_k)$, with $a_i \in A, i = 1, \ldots, k$. Such a path is called a *circuit* if $a_k = a_0$. The *length* of a path is the number of arcs it consists of.

We will also use the graph representation where there exists an arc from a to b when $a\,P\,b$ and there exists an edge between a and b when $a\,I\,b$, as illustrated in figure 3.1.

Another way of representing a binary relation S is to associate, to each element of A, a line and a column of a matrix M^S. The element M^S_{ab} of this matrix (at the intersection of the line associated to a and the column associated to b) is equal to 1 if $a\,S\,b$ and equal to 0 if $a\,\neg S\,b$.

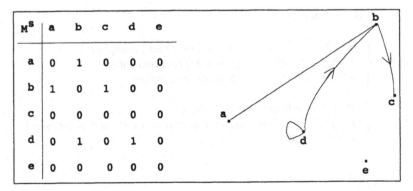

M^S	a	b	c	d	e
a	0	1	0	0	0
b	1	0	1	0	0
c	0	0	0	0	0
d	0	1	0	1	0
e	0	0	0	0	0

Figure 3.1: Graph and matrix representations of a binary relation

3.7   Semiorder

The concept of semiorder was introduced and used in many different contexts, which have been described in chapter 2; the reader has hopefully found there the motivation for introducing this notion and reasons for using it. Many equivalent definitions of a semiorder can be given; some of them will be mentioned in this section, some others will appear in the next chapters, in function of the context. As a starting point, we will consider the following definition (which corresponds to the intuitive definition given in chapter 1).

Definition 3.1 *A reflexive relation $S = (P, I)$ on a finite set A is a semiorder iff there exist a real-valued function g, defined on A, and a nonnegative constant q such that, $\forall a, b \in A$,*

$$(3.4) \qquad \begin{cases} a\,P\,b & \Longleftrightarrow \quad g(a) > g(b) + q, \\ a\,I\,b & \Longleftrightarrow \quad |g(a) - g(b)| \leq q. \end{cases}$$

As illustrated in chapter 2, this definition naturally appears in preference modelling where P denotes "strict preference" and I denotes "indifference", and where a is indifferent to b when their "evaluations" are close enough.

Remark that, due to the connections between S, P and I, it is equivalent to say that S is a semiorder iff there exist a function g and a nonnegative constant q such that, $\forall a, b \in A$,

$$(3.5) \qquad a\,S\,b \Longleftrightarrow g(a) \geq g(b) - q.$$

3.8   Some equivalent definitions of a semiorder

Theorem 3.1 *For a reflexive relation $S = (P, I)$ on a finite set A, the following assertions are equivalent:*

(i) S is a semiorder;

(ii) $\begin{cases} S \cup \overline{S} = A \times A & \text{(S is reflexive and complete)}, \\ S\,S^d\,S \subset S & \text{(S is a Ferrers relation)}, \\ S\,S\,S^d \subset S & \text{(S is semitransitive)}; \end{cases}$

(iii) $\begin{cases} P \cup I \text{ is reflexive and complete}, \\ P\,I\,P \subset P, \text{ (which implies transitivity of } P, \text{ as } I \text{ is reflexive)}, \\ P^2 \cap I^2 = \emptyset; \end{cases}$

(iv)
$$\begin{cases} P \cup I & \text{is} & \text{reflexive and complete,} \\ PIP & \subset & P, \\ P^2 I & \subset & P(\text{ or, equivalently, } I\,P^2 \subset P); \end{cases}$$

(v)
$$\begin{cases} S \text{ is reflexive and complete,} \\ \text{every circuit of length} \leq 4 \text{ in } (A, S) \text{ contains strictly more } I \\ \text{than } P; \end{cases}$$

(vi)
$$\begin{cases} S \text{ is reflexive and complete,} \\ \text{every circuit in } (A, S) \text{ contains strictly more } I \text{ than } P; \end{cases}$$

(vii) there exist a real-valued function g' and a nonnegative function q' defined on A, such that, $\forall a, b \in A$,
$$\begin{cases} a\,P\,b & \Longleftrightarrow & g'(a) > g'(b) + q'(b), \\ a\,I\,b & \Longleftrightarrow & \begin{cases} g'(a) \leq g'(b) + q'(b) \\ g'(b) \leq g'(a) + q'(a); \end{cases} \\ g'(a) > g'(b) & \Longrightarrow & g'(a) + q'(a) \geq g'(b) + q'(b). \end{cases}$$

The proof is given in section 3.25.

Remark that the first property of (ii) is a corollary of the second property and could be deleted. Indeed, if $b\neg Sa$, then $a\,S\,a$, $a\,S^d\,b$, $b\,S\,b$, implying $a\,S\,b$. However, we explicitly maintain the property for the clarity of the proofs.

3.9   Dual of a semiorder, strict semiorder

Theorem 3.2 *A relation S is a semiorder iff its dual $V = S^d$ satisfies the following properties:*

$$\begin{cases} V & \text{is} & \text{irreflexive,} \\ V\,V^d\,V & \subset & V \ (V \text{ is a Ferrers relation}), \\ V\,V\,V^d & \subset & V \ (V \text{ is semitransitive}). \end{cases}$$

The proof is given in section 3.25.

Definition 3.2 *A relation V satisfying the three properties of theorem 3.2 is a strict semiorder.*

Theorem 3.3

    a) If S is a semiorder, then its asymmetric part P is a strict semiorder.

    b) Every strict semiorder is the asymmetric part of exactly one semiorder.

The proof is given in section 3.25.

3.10   Complete preorder

Definition 3.3 *A reflexive relation $S = (P, I)$ on a finite set A is a complete preorder iff there exists a real valued function g, defined on A, such that, $\forall a, b \in A$,*

(3.6)
$$\begin{cases} a\,P\,b & \Longleftrightarrow & g(a) > g(b), \\ a\,I\,b & \Longleftrightarrow & g(a) = g(b). \end{cases}$$

A complete preorder is a semiorder where the constant q is equal to zero; it allows to rank the elements of A from "the best" to "the worse" with eventual ties, in the decreasing order of their values for the function g.

An obvious property is the following: if S is a complete preorder, then I is an equivalence relation.

3.11   Some equivalent definitions of a complete preorder

Theorem 3.4 *For a relation $S = (P, I)$ on a finite set A, the following assertions are equivalent:*

(i)   *S is a complete preorder;*

(ii)   *S is reflexive, complete and transitive;*

(iii) $\begin{cases} P \cup I \text{ is reflexive and complete,} \\ P \text{ is negatively transitive;} \end{cases}$

(iv) $\begin{cases} P \cup I \text{ is reflexive and complete,} \\ P \text{ is transitive,} \\ I \text{ is transitive;} \end{cases}$

(v) $\begin{cases} P \cup I \text{ is reflexive and complete,} \\ P \text{ is transitive,} \\ PI \subset P \ (\text{or, equivalently, } IP \subset P); \end{cases}$

(vi) $\begin{cases} S \text{ is reflexive and complete,} \\ \text{every circuit of length} \leq 3 \text{ in } (A, S) \text{ contains no } P; \end{cases}$

(vii) $\begin{cases} S \text{ is reflexive and complete,} \\ \text{every circuit in } (A, S) \text{ contains no } P. \end{cases}$

The proof is given in section 3.25.

3.12   Dual of a complete preorder, weak order

Theorem 3.5 *A relation S is a complete preorder iff its dual $V = S^d$ is asymmetric and negatively transitive.*

The proof is given in section 3.25.

Definition 3.4 *An asymmetric and negatively transitive relation is called a weak order.*

Theorem 3.6
    a) If S is complete preorder, then its asymmetric part P is a weak order.
    b) Every weak order is the asymmetric part of exactly one complete preorder.

The proof is given in section 3.25.

3.13   Complete order

Definition 3.5 *A reflexive relation $S = (P, I)$ on a finite set A is a complete order iff there exists a real-valued function g, defined on A, such that $\forall a, b \in A$,*

$$(3.7) \qquad \left\{ \begin{array}{lll} a \, P \, b & \Longleftrightarrow & g(a) > g(b), \\ a \neq b & \Longrightarrow & g(a) \neq g(b) \,. \end{array} \right.$$

A complete order is a complete preorder without tie; relation I is reduced to the pairs (a, a) (i.e. to the loops of the graph).

3.14   Some equivalent definitions of a complete order

Theorem 3.7 *For a relation $S = (P, I)$ on a finite set A, the following assertions are equivalent:*

    (i)   *S is a complete order;*
    (ii)   *S is reflexive, complete, antisymmetric and transitive;*
    (iii) $\left\{ \begin{array}{l} P \cup I \text{ is reflexive and complete,} \\ P \text{ is transitive,} \\ I = \{ (a, a), \, \forall a \in A \}; \end{array} \right.$
    (iv)   *P is complete and (A, P) contains no circuits of length ≤ 3;*
    (v)   *P is complete and (A, P) contains no circuit.*

The proof is left to the reader.

3.15   Dual of a complete order, strict complete order

Theorem 3.8 *A relation S is a complete order iff its dual $V = S^d$ is asymmetric, transitive and complete.*

The proof is left to the reader.

Definition 3.6 *An asymmetric, transitive and complete relation is a strict complete order.*

Theorem 3.9

a) *If S is a complete order, then its asymmetric part P is a strict complete order.*

b) *Every strict complete order is the asymmetric part of exactly one complete order.*

The proof is left to the reader.

3.16 Complete preorders and strict complete order associated to a semiorder

Let $S = (P, I)$ be a semiorder and let

$$
\begin{aligned}
S^+(a) &= \{b \in A : aSb\}, \\
S^-(a) &= \{b \in A : bSa\}, \\
P^+(a) &= \{b \in A : aPb\}, \\
P^-(a) &= \{b \in A : bPa\}, \\
d_S^+(a) &= |S^+(a)|, \\
d_S^-(a) &= |S^-(a)|, \\
d_P^+(a) &= |P^+(a)|, \\
d_P^-(a) &= |P^-(a)|, \\
d_S(a) &= d_S^+(a) - d_S^-(a), \\
d_P(a) &= d_P^+(a) - d_P^-(a).
\end{aligned}
$$

Consider the following relations

$$
\begin{aligned}
a\,S_1^+\,b \quad &\text{iff} \quad S^+(b) \subseteq S^+(a), \\
a\,S_2^+\,b \quad &\text{iff} \quad d_S^+(a) \geq d_S^+(b), \\
a\,S_1^-\,b \quad &\text{iff} \quad S^-(a) \subseteq S^-(b), \\
a\,S_2^-\,b \quad &\text{iff} \quad d_S^-(a) \leq d_S^-(b), \\
a\,S^0\,b \quad &\text{iff} \quad d_S(a) \geq d_S(b), \\
a\,P_1^+\,b \quad &\text{iff} \quad P^+(b) \subseteq P^+(a), \\
a\,P_2^+\,b \quad &\text{iff} \quad d_P^+(a) \geq d_P^+(b), \\
a\,P_1^-\,b \quad &\text{iff} \quad P^-(a) \subseteq P^-(b), \\
a\,P_2^-\,b \quad &\text{iff} \quad d_p^-(a) \leq d_P^-(b), \\
a\,P^0\,b \quad &\text{iff} \quad d_P(a) \geq d_P(b), \\
a\,T_1\,b \quad &\text{iff} \quad a\,PI\,b, \\
a\,T_2\,b \quad &\text{iff} \quad a\,IP\,b, \\
a\,T\,b \quad &\text{iff} \quad a\,(PI \cup IP)\,b.
\end{aligned}
$$

It is essential here to remember the second important convention introduced in section 3.4, which excludes to have two distinct elements a and b such that $S^+(a) =$

$S^+(b)$ and $S^-(a) = S^-(b)$ simultaneously. About this point, see also the remark at the end of this section.

Theorem 3.10

a) *If $S = (P, I)$ is a semiorder, then*

  1) $S_1^+ = S_2^+ = P_1^- = P_2^-$ *and they are complete preorders whose asymmetric part is T_2 (which is a weak order);*

  2) $S_1^- = S_2^- = P_1^+ = P_2^+$ *and they are complete preorders whose asymmetric part is T_1 (which is a weak order);*

  3) $S_1^+ \cap S_1^- = S_2^+ \cap S_2^- = P_1^+ \cap P_1^- = P_2^+ \cap P_2^- = S^0 = P^0$ *and they are complete orders whose asymmetric part is $T_1 \cup T_2 = T$ (which is a strict complete order).*

  4) *T is the strict complete order defined by*

$$a T b \iff g(a) > g(b),$$

  where g and q are a numerical representation of P, I according to definition 3.1.

b) *An asymmetric relation P is a strict semiorder iff there is a strict complete order T such that*

$$\begin{cases} a P b, b T c & \implies a P c, \\ a T b, b P c & \implies a P c. \end{cases}$$

c) *A reflexive relation S is a semiorder iff there is a strict complete order T such that*

$$\begin{cases} a S b, b T c & \implies a S c, \\ a T b, b S c & \implies a S c. \end{cases}$$

The proof is given in section 3.25.

Remark

This section shows that with the convention of section 3.4, it is possible to associate a unique strict complete order with every semiorder.

If the convention is not satisfied, $PI \cup IP \cup \{(a, a), a \in A\}$ is a complete preorder; we call it the *complete preorder associated with semiorder (P, I)*. In this case, several strict complete orders are associated with semiorder (P, I); they only differ on the equivalence classes of (P, I). This remark will be used in chapters 5 and 6.

3.17 Graph and matrix representations of a semiorder

It follows from the definition of a semiorder that its graph representation is characterized by the fact that the configurations of figure 3.2 are forbidden, where the "diagonals" are arbitrary and where two indifferent elements can be identical.

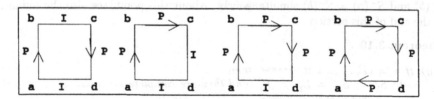

Figure 3.2: Forbidden configurations in a semiorder

It follows from theorem 3.10 b) and c) that if the line and column entries of M^S are ordered according to the strict complete order T, we obtain a *lower-diagonal step-type matrix* as illustrated in the next section. With the same ordering of the lines and columns, M^P will be an *upper-diagonal step-type matrix* which can be easily deduced from M^S : the frontier between 0s and 1s in M^S and M^P are symmetric with respect to the main diagonal (see the example of next section and, in particular, figure 3.3). Reciprocally, every lower-diagonal step-type matrix defines a semiorder and every upper-diagonal step-type matrix defines a strict semiorder (with the condition that the lines and columns are ordered according to the same strict complete order). The general structure corresponding to a step-type matrix is called a *coherent biorder* and has been studied in Doignon, Monjardet, Roubens and Vincke 1986. Semiorders and strict semiorders are respectively reflexive and irreflexive coherent biorders.

3.18   Example of semiorder

Let S be the reflexive relation defined by the following matrix

S	a	b	c	d	e	f	g	h
a	1	1	1	1	1	1	0	1
b	1	1	0	1	0	1	0	1
c	1	1	1	1	1	1	1	1
d	1	1	0	1	1	1	0	1
e	1	1	1	1	1	1	0	1
f	0	1	0	1	0	1	0	0
g	1	1	1	1	1	1	1	1
h	1	1	0	1	1	1	0	1

The next tableau gives the values of d_S^+, d_S^- and d_S (cf. section 3.16).

	a	b	c	d	e	f	g	h
d_S^+	7	5	8	6	7	3	8	6
d_S^-	7	8	4	8	6	8	2	7
d_S	0	-3	4	-2	1	-5	6	-1

The complete preorder S_2^+, based on d_S^+, is the ranking ($\{$ c, g $\}, \{a, e\}$, $\{d, h\}, \{b\}, \{f\}$).

The complete preorder S_2^-, based on d_S^-, is the ranking $(\{g\}, \{c\}, \{e\},$
$\{a, h\}, \{b, d, f\})$.
The complete order S^0, based on d_S, is (g, c, e, a, h, d, b, f).

Ranking the lines and the columns of the matrix according to this order, we obtain step-type matrices presented in figure 3.3, proving that S is a semiorder.

S	g	c	e	a	h	d	b	f
g	1	1	1	1	1	1	1	1
c	1	1	1	1	1	1	1	1
e	0	1	1	1	1	1	1	1
a	0	1	1	1	1	1	1	1
h	0	0	1	1	1	1	1	1
d	0	0	1	1	1	1	1	1
b	0	0	0	1	1	1	1	1
f	0	0	0	0	0	1	1	1

P	g	c	e	a	h	d	b	f
g	0	0	1	1	1	1	1	1
c	0	0	0	0	1	1	1	1
e	0	0	0	0	0	0	1	1
a	0	0	0	0	0	0	0	1
h	0	0	0	0	0	0	0	1
d	0	0	0	0	0	0	0	0
b	0	0	0	0	0	0	0	0
f	0	0	0	0	0	0	0	0

Figure 3.3: Matrix representations of S and P

The matrix of P is obtained from the matrix of S by transposition and replacement of $0's$ by $1's$ and $1's$ by $0's$ ($P = S^d$).

As an exercise, the reader can determine the quantities d_P^+, d_P^-, d_P, the rankings P_2^+, P_2^-, P^0 and verify theorem 3.10. He can also illustrate the properties of theorem 3.1 by using the graph representation of P given in figure 3.4 (the absence of both arcs between two nodes corresponds to an I-edge).

A numerical representation of S can be obtained by defining a function in agreement with the order S^0 and choosing a suitable threshold q' for each element; for instance, we have:

	f	b	d	h	a	e	c	g
g'	0	1	2	3	4	5	6	7
q'	2	3	3	2	2	1	1	0

Another way is to fix a constant threshold q and to adjust the function (below, $q = 1$):

	f	b	d	h	a	e	c	g
	0	0.5	1	1.25	1.5	2	2.5	3.5

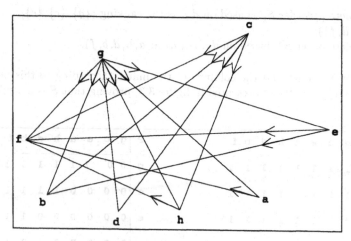

Figure 3.4: Graph representation of P

3.19   Interval order

The concept of interval order generalizes that of semiorder by considering q as a function and not a constant any more. The interested reader will find again in chapter 2 the motivation for introducing this notion.

Definition 3.7 *A reflexive relation $S = (P, I)$ on a finite set A is an interval order iff there exist a real-valued function g and a nonnegative function q such that, $\forall a, b \in A$,*

$$
\begin{cases}
a\,P\,b & \iff \quad g(a) > g(b) + q(g(b)), \\
a\,I\,b & \iff \quad \begin{cases} g(a) \leq g(b) + q(g(b)), \\ g(b) \leq g(a) + q(g(a)), \end{cases}
\end{cases}
$$

3.20   Some equivalent definitions of an interval order

Theorem 3.11 *For a reflexive relation $S = (P, I)$ on a finite set A, the following assertions are equivalent:*

(i)   *S is an interval order;*

(ii) $\begin{cases} S \cup \overline{S} = A \times A \text{ (S is reflexive and complete)}, \\ S\,S^d\,S \subset S \text{ (S is a Ferrers relation)}; \end{cases}$

(iii) $\begin{cases} P \cup I \quad \text{is} \quad \text{reflexive and complete,} \\ PIP \quad \subset \quad P; \end{cases}$

(iv) $\begin{cases} S \text{ is reflexive and complete,} \\ \text{every circuit of length} \leq 4 \text{ in } (A, S) \text{ contains at least} \\ \text{two consecutive } I; \end{cases}$

(v) $\begin{cases} S \text{ is reflexive and complete,} \\ \text{every circuit in } (A, S) \text{ contains at least two consecutive } I. \end{cases}$

The proof is given in section 3.25.

3.21   Dual of an interval order, strict interval order

Theorem 3.12 *A reflexive relation S is an interval order iff its dual $V = S^d$ satisfies the following properties:*

$$\begin{cases} V \text{ is irreflexive,} \\ V\,V^d\,V \subset V \ (V \text{ is a Ferrers relation}). \end{cases}$$

The proof is similar to that of theorem 3.2.

Definition 3.8 *A relation V satisfying the properties of theorem 3.12 is a strict interval order.*

Theorem 3.13
> a) *If S is an interval order, then its asymmetric part P is a strict interval order.*
> b) *Every strict interval order is the asymmetric part of exactly one interval order.*

The proof is similar to that of theorem 3.3.

3.22   Complete preorders and strict complete orders associated to an interval order

Let $S = (P, I)$ be an interval order and let $S_1^+, S_2^+, S_1^-, S_2^-, P_1^+, P_2^+, P_1^-, P_2^-$, T_1 and T_2, the relations defined as in section 3.16. Moreover let

$$\begin{cases} a\,R_1\,b \text{ iff } a\,T_1\,b \text{ or } (a\,\neg T_1\,b, b\,\neg T_1\,a, a\,T_2\,b), \\ a\,R_2\,b \text{ iff } a\,T_2\,b \text{ or } (a\,\neg T_2\,b, b\,\neg T_2\,a, a\,T_1\,b). \end{cases}$$

Theorem 3.14

a) *If* $S = (P, I)$ *is an interval order, then*

    1) $S_1^+ = S_2^+ = P_1^- = P_2^-$ *and they are complete preorders whose asymmetric part is* T_2 *(which is a weak order);*

    2) $S_1^- = S_2^- = P_1^+ = P_2^+$ *and they are complete preorders whose asymmetric part is* T_1 *(which is a weak order);*

    3) R_1 *and* R_2 *are two strict complete orders such that*

$$\begin{cases} a\,P\,b, b\,R_2\,c \Longrightarrow a\,P\,c; \; a\,S\,b, b\,R_1\,c \Longrightarrow a\,S\,c, \\ a\,R_1\,b, b\,P\,c \Longrightarrow a\,P\,c; \; a\,R_2\,b, b\,S\,c \Longrightarrow a\,S\,c. \end{cases}$$

b) *An asymmetric relation* P *is a strict interval order*
iff there is a strict complete order R_1 *such that* $a\,R_1\,b$, $b\,P\,c \Longrightarrow a\,P\,c$
iff there is a strict complete order R_2 *such that* $a\,P\,b$, $b\,R_2\,c \Longrightarrow a\,P\,c$.

c) *A reflexive relation* S *is an interval order*
iff there is a strict complete order R_2 *such that* $a\,R_2\,b$, $b\,S\,c \Longrightarrow a\,S\,c$
iff there is a strict complete order R_1 *such that* $a\,S\,b$, $b\,R_1\,c \Longrightarrow a\,S\,c$.

The proof is given in section 3.25.

3.23   Graph and matrix representations of an interval order

It follows from the definition of an interval order that its graph representation is characterized by the fact that the configurations of figure 3.5 are forbidden, where the "diagonals" are arbitrary and where two indifferent elements can be identical.

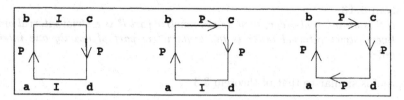

Figure 3.5: Forbidden configurations in an interval order

It follows from theorem 3.14 that if the line and column entries of M^S are respectively ordered according to the strict complete orders R_2 and R_1 (previous section), we obtain a *lower-diagonal step-type matrix* as illustrated in the next section. With the same orderings, but respectively for the columns and the lines, M^P will be an *upper diagonal step-type matrix* (see the example of next section). Reciprocally, every lower-diagonal step-type matrix defines an interval order and every upper-diagonal step-type matrix defines a strict interval order. The general structure corresponding to a step-type matrix is called a *biorder* and has been studied in Doignon et al. 1986. The interval orders and the strict interval orders are respectively reflexive and irreflexive biorders. If the rankings of the lines and columns are the same, the biorder is said to be coherent (see section 3.17).

3.24 Example of interval order

Let S be the reflexive relation defined by the following matrix

S	a	b	c	d	e	f	g	h
a	1	1	1	1	1	1	1	1
b	0	1	0	1	0	1	0	1
c	1	1	1	1	1	1	1	1
d	1	1	1	1	1	1	0	1
e	1	1	0	1	1	1	0	1
f	0	0	0	0	0	1	0	0
g	1	1	1	1	1	1	1	1
h	1	1	1	1	1	1	1	1

The next tableau gives the values of d_S^+, d_S^- and d_S (cf. section 3.16).

	a	b	c	d	e	f	g	h
d_S^+	8	4	8	7	6	1	8	8
d_S^-	6	7	5	7	6	8	4	7
d_S	2	−3	3	0	0	−7	4	1

The complete preorder S_2^+, based on d_S^+, is the ranking ($\{a, c, g, h\}$, $\{d\}$, $\{e\}$, $\{b\}$, $\{f\}$). The complete preorder S_2^-, based on d_S^-, is the ranking ($\{g\}$, $\{c\}$, $\{a, e\}$, $\{b, d, h\}$, $\{f\}$). The strict complete order R_2 is deduced from S_2^+, by ranking the ties on the basis of S_2^-, giving (g, c, a, h, d, e, b, f). The strict complete order R_1 is deduced from S_2^- by ranking the ties on the basis of S_2^+, giving (g, c, a, e, h, d, b, f). Ranking the lines and columns of M^S (resp. the columns and lines of M^P) according to R_2 and R_1 respectively, we obtain the step-type matrices presented in figure 3.6.

The matrix of P is obtained from the matrix of S by transposition and replacement of the $0's$ by $1's$ and $1's$ by $0's$ ($P = S^d$).

As an exercise, the reader can verify the other assertions of theorem 3.14. He can also illustrate the properties of theorem 3.14 by using the graph representation of P given in figure 3.7 (the absence of both arcs between two nodes corresponds to an I-edge).

A numerical representation of S can be obtained by defining a function in agreement with R_1 and choosing a suitable threshold for each element; we have, for instance,

	f	b	d	h	e	a	c	g
Function	0	1	2	3	4	5	6	7
Threshold	0.5	2	4	4	1	2	1	1

S	g	c	a	e	h	d	b	f
g	1	1	1	1	1	1	1	1
c	1	1	1	1	1	1	1	1
a	1	1	1	1	1	1	1	1
h	1	1	1	1	1	1	1	1
d	0	1	1	1	1	1	1	1
e	0	0	1	1	1	1	1	1
b	0	0	0	0	1	1	1	1
f	0	0	0	0	0	0	0	1

P	g	c	a	h	d	e	b	f
g	0	0	0	0	1	1	1	1
c	0	0	0	0	0	1	1	1
a	0	0	0	0	0	0	1	1
e	0	0	0	0	0	0	1	1
h	0	0	0	0	0	0	0	1
d	0	0	0	0	0	0	0	1
b	0	0	0	0	0	0	0	1
f	0	0	0	0	0	0	0	0

Figure 3.6: Matrix representations of S and P

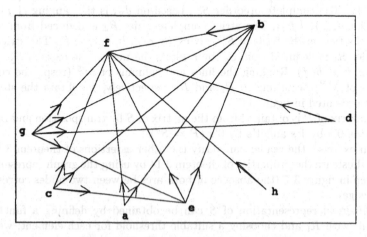

Figure 3.7: Graph representation of P

Remark that it is not a semiorder because, for instance,

$$g \, P \, e \, P \, b \text{ and } g \, I \, h \, I \, b,$$

violating the condition $P^2 \cap I^2 = \emptyset$.

3.25 Proofs of the theorems

Proof of theorem 3.1

$(i) \Longrightarrow (ii)$.

From definition 3.1, we obtain
$$
\begin{array}{rcl}
aSb & \Longleftrightarrow & g(a) \geq g(b) - q, \\
a\bar{S}b & \Longleftrightarrow & g(b) \geq g(a) - q, \\
aS^d b & \Longleftrightarrow & g(b) < g(a) - q,
\end{array}
$$

so that,

- $\forall a, b \in A : aSb$ or bSa (S is reflexive and complete);

- $\forall a, b, c, d \in A : aSb, bS^d c, cSd$ imply $g(a) \geq g(b) - q > g(c) \geq g(d) - q$, giving aSd;

- $\forall a, b, c, d \in A : a \, S \, b, b \, S \, c, c \, S^d \, d$ imply $g(a) \geq g(b) - q \geq g(c) - 2q > g(d) - q$, giving $a \, S \, d$.

$(ii) \Longrightarrow (iii)$.

- $P \cup I$ is reflexive and complete as $S = P \cup I$;

- $a \, P \, b, b \, I \, c, c \, P \, d \Longrightarrow a \, P \, d$ because if $a \, I \, d$ or $d \, P \, a$, then $b \, S \, c$, $c \, S^d \, d$, $d \, S \, a$ would imply $b \, S \, a$, which is not true;

- if $a \, P \, b, b \, P \, c, a \, I \, d, d \, I \, c$, then $c \, S \, d, d \, S \, a, a \, S^d \, b$, giving $c \, S \, b$ which is not true.

$(iii) \Longrightarrow (iv)$.

$a \, P \, b, b \, P \, c, c \, I \, d$ imply $a \neg I \, d$; $d \, P \, a$ would imply $d \, P \, c$ (by transitivity of P), a contradiction.

$(iv) \Longrightarrow (v)$.
Immediate.

$(v) \Longrightarrow (vi)$.
Suppose (A, S) has a circuit C with not more I than P; there exist four elements $a, b, c, d \in C$ such that $a \, I \, b, b \, P \, c, c \, P \, d$, or $a \, P \, b, b \, I \, c, c \, P \, d$; in both cases property (v) implies $a \, P \, d$; we then obtain a shorter circuit (ad replacing $abcd$) containing not more I than P and the same argument works.

$(vi) \Longrightarrow (i)$.
Let us first recall a graph theory result (see Roy 1969): *in a valued graph, there exists a real-valued function g such that $g(a) - g(b) \geq v(a, b)$ iff there is no*

positive valued circuit (where $v(a, b)$ is the value associated to arc (a, b)). Giving value $-q$ to the I type arcs and value $(q + \epsilon)$ to the P type arcs, where q is a positive constant and ϵ, a positive constant such that $m\epsilon < q$ (m being the total number of S arcs), we obtain a valued graph without positive valued circuit (as there are more I than P in every circuit). As a consequence, there exists a real-valued function g such that

$$
\begin{aligned}
a\,P\,b \quad &\Longrightarrow \quad g(a) - g(b) \geq q + \epsilon, \\
a\,I\,b \quad &\Longrightarrow \quad b\,I\,a\,(\text{ by definition of } I), \\
&\Longrightarrow \quad g(a) - g(b) \geq -q \text{ and } g(b) - g(a) \geq -q, \\
&\Longrightarrow \quad |g(a) - g(b)| \leq q.
\end{aligned}
$$

As $P \cup I$ is complete and P and I are mutually exclusive, the converse implications result from the previous ones.

$(i) \Longrightarrow (vii)$.
Obvious.

$(vii) \Longrightarrow (ii)$.
Similar to $(i) \Longrightarrow (ii)$.

Proof of theorem 3.2

- S is reflexive iff V is irreflexive by the definition of the dual;

- let $a\,V\,b, b\,V^d\,c, c\,V\,d$ and $a\,\neg V\,d$; then $b\,\neg S\,a, bSc, d\,\neg S\,c$ and dSa, so that $bScS^ddSa$ which implies bSa, a contradiction. Permuting S with V, we prove that $V\,V^d\,V \subset V$ implies $S\,S^d\,S \subset S$;

- let $a\,V\,b, b\,V\,c, c\,V^d\,d$ and $a\,\neg V\,d$; then $b\,\neg S\,a, c\,\neg S\,b, cSd$ and $d\,S\,a$, so that $c\,S\,d\,S\,a\,S^d\,b$, which implies $c\,S\,b$ a contradiction. Permuting S with V, we prove that $V\,V\,V^d \subset V$ implies $S\,S\,S^d \subset S$.

Proof of theorem 3.3

a) If S is semiorder, then $a\,P\,b \Longleftrightarrow b\,\neg S\,a \Longleftrightarrow a\,S^d\,b$, so that P satisfies the properties of theorem 3.2.
b) Let V be a strict semiorder and $S = V^d$. By theorem 3.2, S is a semiorder; moreover $a\,V\,b \Longleftrightarrow a\,S^d\,b \Longleftrightarrow b\,\neg S\,a$ and $a\,Sb$ (S is complete), so that V is the asymmetric part of S.

Proof of theorem 3.4

$(i) \Longrightarrow (ii)$.
Results from the fact that $a\,S\,b \Longleftrightarrow g(a) \geq g(b)$.

$(ii) \implies (iii)$.
$a \neg P b, b \neg P c \implies a \neg S b, b \neg S c \implies b S a, c S b \implies c S a \implies a \neg P c.$

$(iii) \implies (iv)$.
$a P b, b P c \implies a P c$; otherwise, $a \neg P c, c \neg P b$ would imply $a \neg P b$;
transitivity of I results from negative transitivity of P and from the fact that $a I b$
iff $a \neg P b, b \neg P a$.

$(iv) \implies (v)$.
$a P b, b I c \implies a P c$ because $c I a$, with $b I c$, would imply $a I b$, and $c P a$, with
$a P b$, would imply $c P b$.

$(v) \implies (vi)$.
Immediate.

$(vi) \implies (vii)$.
Suppose (A, S) has a circuit C with a P; there exist three elements $a, b, c \in C$
such that $a P b, b I c$ or $a P b, b P c$, giving a shorter circuit with $a P c$, and the same
argument works.

$(vii) \implies (i)$.
Same proof as for the implication $(vi) \implies (i)$ of theorem 3.1, with $q = 0$.

Proof of theorem 3.5

As $V = S^d = P$, the asymmetry and negative transitivity of V result from the
properties of P (see theorem 3.4).
    If $V = S^d$ is asymmetric and negatively transitive, then S is reflexive ($a S a$
because $a \neg V a$), complete ($a \neg S b \implies b V a \implies a \neg V b \implies b S a$) and transitive
($a S b$ and $b S c \implies b \neg V a$ and $c \neg V b \implies c \neg V a$
$\implies a S c$).

Proof of theorem 3.6

Similar to the proof of theorem 3.3.

Proof of theorem 3.10

a.1) By definition, S_2^+ is a complete preorder (see section 3.10).
    $a S_2^+ b \implies a S_1^+ b$:
    if not, $\exists c : b S c$ and $c S^d a$; due to the fact that $d_S^+(a) \geq d_S^+(b)$, this implies
    the existence of $d : a S d$ and $b \neg S d$; but $b S c, c S^d a, a S d$ imply $b S d$ (theorem
    3.1), a contradiction;
    $a S_1^+ b \implies a P_1^- b$ because $S^+(a) \cup P^-(a) = S^+(b) \cup P^-(b) = A$;
    $a P_1^- b \implies a P_2^- b$: trivial;
    $a P_2^- b \implies a S_2^+ b$ because $d_S^+(a) + d_P^-(a) = d_S^+(b) + d_P^-(b) = |A|$. This com-
    pletes the proof of $S_2^+ = S_1^+ = P_1^- = P_2^-$.

If $aS_1^+b, b\neg S_1^+a, \exists d : aSd, dPb \implies a(I \cup P)dPb \implies aT_2b$; if $aIPb$ and bSc; then aSc; if not, $cPaIPb$ gives cPb, a contradiction; this proves that $aT_2b \implies aS_1^+b$.

a.2) Similar to a.1).

a.3) The second convention of section 3.4 gives

$$a\neg Eb \implies \exists c : [(aSc, b\neg Sc) \text{ or } (cSb, c\neg Sa)] \text{ or } [(a\neg Sc, bSc) \text{ or } (cSa, c\neg Sb)] \text{ (\underline{or} is exclusive)}$$
$$\implies d_S(a) \neq d_S(b)$$
$$\implies S^0 \text{ is a complete order (see section 3.13).}$$

We prove now that the asymmetric part of S^0 is T (the proofs of the other equalities are left to the reader) :

$$aS^0b, b\neg S^0a \iff \exists c : (aSc, cPb) \text{ or } (cSb, aPc)$$
$$\iff a(I \cup P)cPb \text{ or } aPc(I \cup P)b$$
$$\iff a(IP \cup PI)b$$
$$\iff aTb.$$

Finally, given a and b with $a \neq b$,

$$aTb \implies \exists c : (aPc \text{ and } b\neg Pc) \text{ or } (cPb \text{ and } c\neg Pa)$$
$$\implies \exists c : g(a) > g(c) + q \geq g(b)$$
$$\text{or } g(a) + q \geq g(c) > g(b) + q$$
$$\implies g(a) > g(b).$$

Conversely $g(a) > g(b) \implies aTb$ (as T is a strict complete order and bTa would imply the converse inequality).

b) From the latter implications of a.3), it immediately follows that, given a strict semiorder P, the associated strict complete order T is such that

$$\begin{cases} aPb, bTc & \implies & aPc, \\ aTb, bPc & \implies & aPc. \end{cases}$$

Conversely, let P be an asymmetric relation and T a strict complete order satisfying the above implications. We have to prove that (cf. definition 3.2)

$$\begin{cases} PP^dP \subset P, \\ PPP^d \subset P . \end{cases}$$

If $aPb, c\neg Pb$ and cPd, then $a \neq c$; cTa would imply $cTaPb$, hence cPb which is not true; thus, aTc which, with cPd, implies aPd. If $aPb, bPc, d\neg Pc$, then $b \neq d$; dTb would imply $dTbPc$, hence dPc which is not true; thus, bTd which, with aPb, implies aPd.

c) Left to the reader (similar to b)).

Proof of theorem 3.11

$(i) \Longrightarrow (ii) \Longrightarrow (iii)$.
See the similar implications in theorem 3.1.

$(iii) \Longrightarrow (iv)$.
Immediate.

$(iv) \Longrightarrow (v)$.
Suppose (A, S) has a circuit C with no two consecutive I; there exist four elements $a, b, c, d, \in C$ such that aPb, bIc, cPd, implying aPd and we obtain a shorter circuit with no two consecutive I where the same argument works.

$(v) \Longrightarrow (ii)$.
Immediate.

$(iii) \Longrightarrow (i)$.
Let

$$aT_1b \text{ iff } a \ P \ I \ b,$$
$$aT_2b \text{ iff } a \ I \ P \ b,$$
$$aR_1b \text{ iff } aT_1b \text{ or } (a\neg T_1b, \ b\neg T_1a, aT_2b).$$

The reader can prove, as an exercise, that R_1 is a strict complete order so that there is a real-valued function g such that

$$aR_1b \Longleftrightarrow g(a) > g(b).$$

Defining

$$g(a) + q(g(a)) = \max\{g(c) : cIa\} \geq g(a) \text{ (reflexivity of I)},$$

we obtain

$$
\begin{aligned}
aPb \ &\Longrightarrow \ aR_1c, \forall c : cIb \\
&\Longrightarrow \ g(a) > g(c), \forall c : cIb \\
&\Longrightarrow \ g(a) > g(b) + q(g(b)); \\
aIb \ &\Longrightarrow \ \begin{cases} g(b) \leq g(a) + q(g(a)), \\ g(a) \leq g(b) + q(g(b)). \end{cases}
\end{aligned}
$$

Proof of theorem 3.14

The proofs of a.1) and a.2) are similar to those of a.1) and a.2) in theorem 3.10.

a.3) From the two first assertions of the theorem, we obtain

$$aR_1b \text{ iff } d_S^-(a) < d_S^-(b) \text{ or } d_S^-(a) = d_S^-(b) \text{ and } d_S^+(a) > d_S^+(b),$$
$$aR_2b \text{ iff } d_S^+(a) > d_S^+(b) \text{ or } d_S^+(a) = d_S^+(b) \text{ and } d_S^-(a) < d_S^-(b),$$

so that R_1 and R_2 are clearly asymmetric and transitive. They are also complete because of our convention of section 3.4 which excludes having $a \neg T_1 b$, $b \neg T_1 a$, $a \neg T_2 b$, $b \neg T_2 a$. Now, if aPb and $bR_2 c$, then aPc because $a \neg Pc$ would imply $c(P \cup I)a \implies cIPb \implies cT_2 b \implies cR_2 b$ which is incompatible with $bR_2 c$. If aSb and $bR_1 c$, then aSc because $a \neg Sc \implies cPa$ which, with aSb, implies $cPIb \implies cT_1 b \implies cR_1 b$ which is incompatible with $bR_1 c$. The other implications are proved in the same way.

b) The first part of the proof is an immediate consequence of the point $a.3)$ above. Conversely let P be an asymmetric relation and R_1 a strict complete order such that

$$aR_1 b, bPc \implies aPc.$$

We have to prove that $PP^d P \subset P$. If $aPb, c \neg Pb, cPd$, then $c \neg R_1 a$ (because $cR_1 aPb \implies cPb$); hence $aR_1 c$, which, with cPd, implies aPd.
The reasoning is similar for R_2.

c) Left to the reader (similar to b)).

4

MINIMAL REPRESENTATIONS

In this chapter, we study in detail some particular numerical representations of a semiorder which are in some sense minimal or parsimonious. Such representations generalize to semiorders the rank of the elements in a strict complete order which means that they have similar rights to pretend at being a "default" numerical representation. The notion of minimal representation of a semiorder was first introduced and proven to exist by Pirlot. The exposition below is essentially based on Doignon 1988, Doignon and Falmagne 1994, Pirlot 1990, Pirlot 1991 and Mitas 1994. Related results have recently and independently be obtained in Troxell 1995 for the symmetric part of a semiorder, i.e. a unit interval graph or indifference graph.

4.1 Potential functions in valued graphs

A useful tool for studying the numerical representations of semiorders consists in representing the inequalities that the pair (g, q) must satisfy (see definition 3.1) by means of a valued graph using the same conventions as in the modelling of project scheduling problems (PERT, Critical Path Method, see e.g. Hillier and Lieberman 1990, pp. 369 sq. and pp. 378-379, for a linear programming formulation).

Let $G = (A, U, v)$ be a valued graph on a finite set of nodes A; a real value $v(a, b)$ is attached to each arc (a, b) of U. As we will see, the function g, in any numerical representation of a semiorder, can be considered as a "potential function" of a suitable valued graph .

Definition 4.1 *A potential function of the valued graph* $G = (A, U, v)$ *is a function* $g : A \to R$ *such that,* $\forall (a, b) \in U$, $g(a) \geq g(b) + v(a, b)$.

It is easily seen that if g is a potential function whose minimal value is 0, then $g(a)$ cannot be smaller than the maximal value of the paths starting from a (the value of a path being the sum of the values of the arcs of the path). A fundamental result is the following (Roy 1969).

71

Theorem 4.1 *A valued graph admits potential functions iff there is no circuit of strictly positive value in the graph. The smallest non-negative potential function assigns to each node the maximal value of the paths starting from the node.*

4.2 Alternative definition of a finite semiorder

There is a small difficulty for representing the strict inequalities of definition 3.1. However, as the set A is finite, it is clear that the following definition is equivalent to definition 3.1.

Definition 4.2 *A reflexive relation $S = (P, I)$ on a finite set A is a semiorder iff there exist a real function g, defined on A, a nonnegative constant q and a positive constant ε such that, $\forall a, b \in A$:*

$$(4.1) \qquad \begin{cases} a\,P\,b & \Longleftrightarrow & g(a) \geq g(b) + q + \varepsilon, \\ a\,I\,b & \Longleftrightarrow & |g(a) - g(b)| \leq q. \end{cases}$$

A triple (g, q, ε) satisfying (4.1) is called an ε-*representation* of (P, I). Remark that any representation (g, q) in the sense of definition 3.1 yields an ε-representation, where

$$\varepsilon = \min_{(a,b) \in P} \{ g(a) - g(b) - q \}.$$

Moreover, the set of all ε'-representations, for $\varepsilon' = \beta\varepsilon$, β being any positive number, is easily obtained from the set of ε-representations; indeed, (g, q, ε) is an ε-representation iff $(\beta g, \beta q, \beta\varepsilon)$ is an ε'-representation. This means that it is sufficient to study the set of ε-representations for fixed ε.

4.3 Valued graph associated with a semiorder

Let (A, S) be the graph associated with the semiorder $S = (P, I)$ on the finite set A as defined in section 3.6. A valued graph denoted $G(q, \varepsilon)$ is obtained by giving the value $(q + \varepsilon)$ to the arcs P and $(-q)$ to the arcs I, the loops being excluded, as illustrated in figure 4.1.

It is clear that (g, q, ε) is an ε-representation of (P, I) iff g is a potential function of $G(q, \varepsilon)$.

This formulation makes it possible to give a precise condition for the existence of a representation with a given threshold q. The reader will remark that the condition only involves the non-valued graph (A, S).

Theorem 4.2 *If $S = (P, I)$ is a semiorder on the finite set A, there exists an ε-representation with threshold q iff*

$$(4.2) \qquad \frac{q}{\varepsilon} \geq \alpha = \max_{C} \left\{ \frac{|C \cap P|}{|C \cap I| - |C \cap P|}, C \text{ circuit of } (A, S) \right\},$$

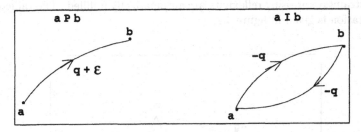

Figure 4.1: Representing a semiorder by a valued graph

where $|C \cap P|$ (resp. $|C \cap I|$) denotes the number of arcs P (resp I) in the circuit C of the graph (A, S).

Proof can be found in section 4.16.

Remark

In the particular case of complete preorders, $|C \cap P| = 0$, for all circuit C and $\alpha = 0$.

4.4 An algorithm for finding the numerical representation of a semiorder

It is immediate from the previous section that one can obtain a numerical representation of a semiorder by the following procedure:

- choose any value for ε, e.g. $\varepsilon = 1$,
- choose a large enough value of $\frac{q}{\varepsilon}$ (e.g. $\frac{q}{\varepsilon} = |P|$),
- solve the maximal value path problem in the graph $G(q, \varepsilon)$ (e.g. by using the Bellman algorithm, see Lawler 1976, pp.74 sq., see also Ahuja, Magnanti and Orlin 1993, pp. 133 sq.).

4.5 Example

Let $S = (P, I)$ be the semiorder on $A = \{a, b, c\}$ defined by $P = \{(a, c)\}$. The constraints (4.1) translate into

$$
\begin{aligned}
g(a) &\geq g(c) + q + \varepsilon & \text{(for } a \, P \, c) \\
\left.\begin{aligned}
g(a) &\geq g(b) - q \\
g(b) &\geq g(a) - q
\end{aligned}\right\} & & \text{(for } a \, I \, b) \\
\left.\begin{aligned}
g(b) &\geq g(c) - q \\
g(c) &\geq g(b) - q
\end{aligned}\right\} & & \text{(for } b \, I \, c)
\end{aligned}
$$

The constraints expressing reflexivity are automatically fulfilled. The valued graph representation is given in figure 4.2.

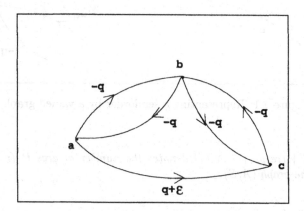

Figure 4.2: Valued graph representation of a semiorder on three points

In this graph, the value of the only non-trivial simple circuit $C = \{(a,c)(c,b)(b,a)\}$ is $-q + \varepsilon$. Hence, the necessary and sufficient condition for the existence of an ε-representation with threshold q is $q \geq \varepsilon$. Various ε-representations with $\varepsilon = 1$ are provided in table 4.1.

		a	b	c
$q = 1$	$g_1 =$	2	1	0
	$g_2 =$	9.5	8.5	7.5
	$g_3 =$	3.5	1	0
$q = 2.5$	$g_4 =$	10.5	8.5	7
	$g_5 =$	3.5	2.5	0

Table 4.1: Various ε-representations with $\varepsilon = 1$

Remark that g_1 and g_3 are the solutions of the maximal value path problem in $G(1,1)$ and $G(2.5,1)$ respectively; g_1 is a very special representation: it is the smallest one among all ε-representations with $\varepsilon = 1$ and has remarkable properties that will be studied in detail in the next section.

4.6 Minimal representation of a semiorder

In this section, we only consider nonnegative representations, which is not restrictive in the case of finite semiorders.

Let us denote by $g_{q,\varepsilon}$ the solution of the maximal path problem in $G(q,\varepsilon)$. By theorem 4.1, for any nonnegative ε-representation g with threshold q, we have, $\forall a \in A$,

$$g_{q,\varepsilon}(a) \leq g(a).$$

A natural question is whether there exists a minimal nonnegative representation among all ε-representations whatever the threshold q is (with $\frac{q}{\varepsilon} \geq \alpha$).

Definition 4.3 *A representation (g^*, q^*, ε) is minimal in the set of all nonnegative ε-representations (g, q, ε) of a semiorder iff $\forall a \in A$, $g^*(a) \leq g(a)$.*

One should call such a representation "ε-minimal" but we drop the ε because if the ε-minimal representation g^* does exist, then one obtains the ε'-minimal representation, for any other $\varepsilon' > 0$, by multiplying g^* by $\varepsilon' / \varepsilon$.

Let $q^* = \alpha\varepsilon$ with α defined as in theorem 4.2. The next result characterizes the minimal representation of a semiorder.

Theorem 4.3 *The representation $(g_{q^*,\varepsilon}, q^*, \varepsilon)$ is minimal in the set of all ε-representations of a semiorder S.*

Proof of theorem 4.3

We have to prove that

$$g_{q,\varepsilon}(a) \geq g_{q^*,\varepsilon}(a),$$

for any $a \in A$ and $q > q^*$ ($q < q^*$ is impossible by theorem 4.2). For any maximal value path Γ starting from a in $G(q,\varepsilon)$, the value $v(\Gamma)$ of Γ is equal to $g_{q,\varepsilon}(a)$ and also

$$v(\Gamma) = (|\Gamma \cap P| - |\Gamma \cap I|)q + |\Gamma \cap P|\varepsilon.$$

The same path, in $G(q^*, \varepsilon)$, will be maximal, with a value equal to $g_{q^*,\varepsilon}(a)$ given by

$$v^*(\Gamma) = (|\Gamma \cap P| - |\Gamma \cap I|)q^* + |\Gamma \cap P|\varepsilon.$$

The theorem directly results from the following proposition in Doignon 1988.

Theorem 4.4 *For any q such that $\frac{q}{\varepsilon} \geq \alpha$, among all paths of maximal value starting from a in $G(q,\varepsilon)$, there is a path Γ for which the number of P-edges is not smaller than the number of I-edges, i.e. $|\Gamma \cap P| \geq |\Gamma \cap I|$, and this, $\forall a \in A$.*

Proof of this theorem can be found in section 4.16. A slightly stronger result is lemma 2.8 in Pirlot 1990.

4.7 Integrality of the minimal representation

A very important property of the minimal representation is that when $\varepsilon = 1$, g^* and $q^* = \alpha$ take integer values. The proof is given in section 4.16.

Theorem 4.5 *In the minimal representation* (g^*, q^*, ε) *of a semiorder,* q^* *and all values of* g^* *are integral multiples of* ε.

As a corollary of this "integrality property", we can state in a loose way that the minimal representation is the smallest representation on the integers and it can be viewed as a generalization of the *rank function* usually associated with strict complete orders. The rank of an element $a \in A$ in a strict complete order can be defined as the number of elements which are "worse" than a, i.e. $|\{x \in A : aPx\}|$. The following definition is in agreement with the usual notion of rank when S is a complete order.

Definition 4.4 *The rank function* r *associated with a semiorder* S *is the smallest numerical representation of* S *in the set of positive integers.*

By theorems 4.3 and 4.5, r is the minimal ε-representation of S with $\varepsilon = 1$. Hence the rank of $a \in A$ is the maximal value of the paths issued from a in the graph graph $G(\alpha\varepsilon, \varepsilon)$ associated with S.

4.8   Maximal contrast property

Another interesting property of the minimal representation can be stated in terms of "contrast" in the following manner: in the minimal representation, the difference between pairs of alternatives which are indifferent and pairs of alternatives which are preferred to one another is maximized. More precisely, consider a constant threshold representation g of a semiorder $S = (P, I)$. Usually the threshold for which g is a representation is not uniquely determined. Let us choose a particular one as follows. Let

$$q_g = \max\{|g(a) - g(b)|,\ aIb\}$$

and

$$q'_g = \min\{g(a) - g(b),\ aPb\}.$$

We define the *contrast* of g and denote ε_g, the difference $q'_g - q_g$; obviously $(g, q_\varepsilon, \varepsilon_g)$ is an ε_g-representation of S.

Theorem 4.6 *Let* g^* *be the minimal* ε-*representation of a semiorder* S. *The contrast of* g^* *is* ε *and* g^* *has maximal contrast among all representations which range in the same interval as* g^*.

The proof is in section 4.16. Let us just state the theorem in another manner to make it more explicit. Let g be any representation of S with range equal to $[0, K]$, i.e. the minimal (resp. maximal) value taken by g is 0 (resp. K). We can associate the values q_g, q'_g and the contrast ε_g as defined above. Let g^* be the minimal ($\varepsilon = 1$)-representation of S; $q^* = \alpha$ is the value of q_{g^*}. and let $\overline{g}^*$ denote the maximal value of g^* on A. Clearly, the function $\widetilde{g}^* = (K/\overline{g}^*)\, g^*$ is a representation of S with range $[0, K]$. The theorem actually tells us that the contrast of this representation is at least as large as the contrast of g.

4.9 An example and the synthetic graph of a semiorder

At this stage, an example of a semiorder together with a few numerical representations including the minimal one would be welcome. Let semiorder S be given by its matrix M^S represented in figure 4.3 where the nodes are ranked and labelled in accordance with the strict complete order T defined in section 3.16.

	1	2	3	4	5	6	7	8	9
1	0	0	1	1	1	1	1	1	1
2	0	0	0	1	1	1	1	1	1
3	0	0	0	0	1	1	1	1	1
4	0	0	0	0	1	1	1	1	1
5	0	0	0	0	0	0	0	1	1
6	0	0	0	0	0	0	0	0	1
7	0	0	0	0	0	0	0	0	0
8	0	0	0	0	0	0	0	0	0
9	0	0	0	0	0	0	0	0	0

Figure 4.3: Example of semiorder given by its matrix

In order to determine representations of S, we have to study the potential functions of the graph $G(q,\varepsilon)$ which is a complete graph on 9 nodes. It is clear that a more synthetic graph representation is highly desirable not only for graphical readability but also for computational reasons. In this section we show informally, on the example, how a synthetic representation can be found; more formal definitions and results are deferred to the next section.

Much in the spirit of the Hasse diagram for transitive relations, the observation of matrix M^S suggests a more concise representation than the graph $G(q,\varepsilon)$. Given the strict complete order T, the matrix is completely known as soon as the positions of its *steps* are known.

Let us call *noses*, the noses of the steps i.e., in the example, the oriented pairs that constitute relation N included in P,

$$N = \{(1,3),(2,4),(4,5),(5,8),(6,9)\}.$$

Similarly, we call *hollows* the symmetric w.r.t. the matrix diagonal of the hollows of the steps, i.e. in the example, the oriented pairs constituting following relation denoted by H and included in I,

$$H = \{(2,1),(3,2),(4,3),(7,5),(8,6),(9,7)\}.$$

Knowing N and/or H, together with T, allows to reconstruct M^S and hence S. Moreover, it is not hard to convince oneself that the system of constraints (4.1) an ε-representation (g,q,ε) has to satisfy, can be summarized by, $\forall a,b \in A$,

(4.3) $a\,T\,b \Longrightarrow g(a) \geq g(b)$,

(4.4) $a\,N\,b \Longrightarrow g(a) \geq g(b) + q + \varepsilon$,

(4.5) $b\,H\,a \Longrightarrow g(a) \leq g(b) + q$.

Indeed, the constraints involving T and N ensure that, for all c, d with $c\,T\,a$ and $b\,T\,d$, $a\,N\,b$ implies

$$g(c) \geq g(d) + q + \varepsilon$$

and hence $c\,P\,d$. More informally, T allows to rank the elements of A and N locates the steps of the semiorder matrix.

Similar reasoning can be done with T and H. As a result, the graph $G(q, \varepsilon)$ can be simplified by representing only the constraints (4.3), (4.4) and (4.5). It is even not necessary to represent all constraints involving T: it is sufficient to represent $a\,T\,b$ when b is ranked immediately after a in order T, which we shall denote by $b = a + 1$. Formally, $b = (a + 1)$ iff $a\,T\,b$ and $\forall x \in A$, with $x \neq a$, $(a\,T\,x \Longrightarrow b\,T\,x)$. Denoting by LA the last element of A in order T (i.e. $\forall x \in A$, $x\,T\,(LA)$), we can replace (4.3) by: $\forall a \in A$ with $a \neq LA$,

(4.6) $g(a) \geq g(a + 1)$.

We call *synthetic graph* and denote by $SG(q, \varepsilon)$ the valued graph representing the constraints (4.6), (4.4) and (4.5) with the same conventions that were used for $G(q, \varepsilon)$.

The synthetic graph associated with our example is illustrated in figure 4.4

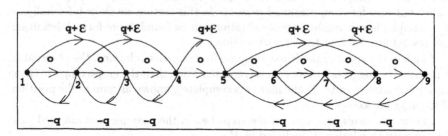

Figure 4.4: Synthetic graph of the semiorder defined in figure 4.3

In this example, the circuit responsible for the determination of α goes through nodes $5, 8, 6, 9, 7, 5$. Its value is

$$2\,(q + \varepsilon) - 3\,q = -q + 2\varepsilon.$$

Hence, q must be at least 2ε and $\alpha = 2$. In this example, it is possible to provide formulae giving the values $g_{q,\varepsilon}(x)$ as affine functions of q and ε $(q \geq 2\varepsilon)$ with appropriate coefficients. It is in fact always possible to write such formulae; this is a consequence of theorem 4.20 below. We give such formulae in the first line of

	1	2	3	4	5	6	7	8	9
$g_{q,\varepsilon}$	$3q + 5\varepsilon$	$3q + 4\varepsilon$	$2q + 4\varepsilon$	$2q + 3\varepsilon$	$q + 2\varepsilon$	$q + \varepsilon$	2ε	ε	0
$g^* = g_{2\varepsilon,\varepsilon}$	11ε	10ε	8ε	7ε	4ε	3ε	2ε	ε	0
$g_{3\varepsilon,\varepsilon}$	14ε	13ε	10ε	9ε	5ε	4ε	2ε	ε	0

Table 4.2: Some numerical representations of the semiorder defined in figure 4.3

table 4.2. The second line is the minimal representation g^* which ranges in the interval $[0, 11\varepsilon]$. There is no other representation in integral multiples of ε in the same interval. The last line gives $g_{3\varepsilon,\varepsilon}$.

In the next section, we will show that the graph associated with the noses and hollows, or equivalently associated with constraints (4.4) and (4.5), completely determines the semiorder (provided there are no equivalent elements).

We call this valued graph the *Super Synthetic Graph* and denote it $SSG(q, \varepsilon)$ or, more concisely, SSG. The SSG associated to the example of this section is represented in figure 4.5. The non-valued version of this graph is simply the graph (A, N, H), a graph on the set A with two types of arcs N and H.

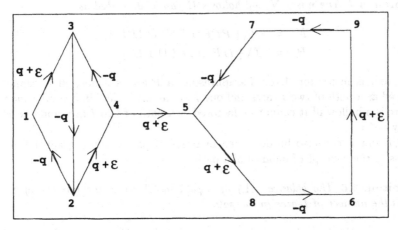

Figure 4.5: Super synthetic graph of the semiorder defined in figure 4.3

4.10 Noses and Hollows

In this section, we give formal definitions of the two relations N ("noses") and H ("hollows") previously introduced. We state properties showing that these relations provide a synthetic but complete description of a semiorder.

Definition 4.5 *Let $P^+(a)$ denote the set of successors of any node $a \in A$ in P, i.e., $P^+(a) = \{\, x \in A, \, a\,P\,x \,\}$ and let $P^-(a)$ denote the set of predecessors of $a \in A$ in P, i.e. $P^-(a) = \{\, y \in A, \, y\,P\,a \,\}$. The pair $(a,b) \in A \times A$ is a nose, $(a,b) \in N$, if*

$$b = \min_T P^+(a) \quad \text{and} \quad a = \max_T P^-(b),$$

where $\min_T$ (resp. $\max_T$) means that the minimal (resp. maximal) element w.r.t. the strict complete order T is considered.

The pair $(b,a) \in A \times A$ is a hollow, $(b,a) \in H$, if

$$b = \min_T [\, A \setminus P^+(a) \,] \quad \text{and} \quad a = \max_T [\, A \setminus P^-(b) \,].$$

The operators " $\min_T$ " and " $\max_T$ " yield the smallest and the greatest nodes w.r.t. T in the subset on which they operate.

Recall that $S = (P, I)$ is supposed to be a semiorder without equivalent elements. From the definition it is clear that $N \subseteq P$ and $H \subseteq I$.

We now give a direct characterization of N and H in terms of P and I; conversely we show how N and H allow to reconstruct S.

Theorem 4.7 *The noses N and hollows H can be described as*

$$N = P \setminus \{\, PPI \cup PIP \cup IPP \,\},$$
$$H = I \setminus \{\, IIP \cup IPI \cup PII \,\}.$$

The proof is in section 4.16. Paraphrasing, a P-arc is a nose iff it cannot be factored as a path of two P-arcs and one I-arc in any order. In a similar manner, a I-arc is a hollow iff it cannot be factored as a path of two I-arcs and one P-arc in any order.

P, I and T can also be described in terms of paths of the graph (A, N, H), thanks to the concept of balance of a path.

Definition 4.6 *The balance $\beta(\Gamma)$ of a path in (A, S) is the number of P-arcs minus the number of I-arcs in the path.*

The next theorem characterizes some paths of (A, S) as paths of (A, N, H).

Theorem 4.8 *Let Γ be a path of maximal balance from a to b in (A, S). Suppose in addition that the number of P-arcs of Γ is maximal in the set of paths of maximal balance from a to b. Then Γ is a simple path of (A, N, H).*

An immediate consequence of theorem 4.8 is that the graph (A, N, H) is just as connected as (A, S). Indeed it is clear that there is path from a to b in (A, N, H) as soon as there is one in (A, S). The only case where there could be no path from a to b in (A, N, H) is when $b\,P\,a$. This will be examined in more detail by the end of this section.

Theorem 4.9 *For any pair a, b in A, with $a \neq b$, we have*

(i) $a\,P\,b$ iff there is a path Γ from a to b in (A, N, H) with $\beta(\Gamma) > 0$;

(ii) $a\,T\,b$ iff there is a path Γ from a to b in (A, N, H) with $\beta(\Gamma) \geq 0$;

(iii) $a\,I\,b$ iff for any path Γ from a to b or from b to a in (A, N, H), $\beta(\Gamma)) \leq 0$.

We can prove a result which is slightly stronger than the third assertion of theorem 4.9.

Theorem 4.10 *For any pair a, b in A, we have $a\,I\,b$ iff the paths of maximal balance from a to b or from b to a in (A, N, H) have balance 0. Moreover, the only case where there are simultaneously a path from a to b and a path from b to a both with balance 0 is when $a = b$ and both paths are empty.*

From previous theorems, we see that the graphs (A, S) and (A, N, H) both provide a complete description of a semiorder. Let us stress the fact that the characterization via (A, N, H) is in terms of structure of paths in the graph and not in terms of valued graphs and potential theory. Note also that the graphs (A, N, T) and (A, H, T) both allow to reconstruct S as they allow to restore the step matrix M^S. Hence, any two of the three relations N, H, T allow to reconstruct a semiorder S. For the pair (N, H), we have given an explicit description of the manner P, T and I can be identified in terms of paths of the graph (A, N, H). Such a precise description was not given for the pairs (T, N) and (T, H).

Let us now turn back to valued graphs and potential functions. The following important result parallels theorem 4.8 for valued graphs.

Theorem 4.11 *All paths of maximal value in $G(q, \varepsilon)$ are paths of $SSG(q, \varepsilon)$ i.e. their arcs either belong to N or to H.*

This result has important consequences. In particular, the systems of constraints (4.4) and (4.5) respectively associated with the arcs of N and H suffice to determine S.

Theorem 4.12 *A triple (g, q, ε) is an ε-representation of semiorder S iff g is a potential function of the valued graph $SSG(q, \varepsilon)$.*

The proof is immediate since all maximal value paths of $G(q, \varepsilon)$ are paths of $SSG(q, \varepsilon)$ according to theorem 4.11.

To conclude this section we investigate the connectivity properties of (A, N, H). We already noticed that (A, N, H) is a connected graph, i.e. for all $a, b \in A$ with $a \neq b$, there is a path from a to b in (A, N, H). We know that it is the case when $(a, b) \in S$ The only case in which there could be no path from a to b in (A, N, H) is the case where $(b, a) \in P$. Consider the elements of A and rank them according to the order T, i.e.

$$x_1\,T\,x_2\,T\,x_3.....T\,x_n.$$

If for all i, we have $x_i \, I \, x_{i+1}$, then there is always a path going from any a to any b using arcs of I and hence a path of (A, N, H). Otherwise suppose that for some $i = i^*$, $x_{i^*} \, P \, x_{i^*+1}$; then, $\forall j \leq i^*$ and $\forall k \geq i^* + 1$, $x_j \, P \, x_k$ and there is no path in (A, S) from the set $\{ \, x_k, \, k \geq i^* + 1 \, \}$ to the set $\{ \, x_j, \, j \leq i^* \, \}$. This implies that in case $(b, a) \in P$, there is a path from a to b in (A, S) (and hence in (A, N, H)) iff for all pairs (x_i, x_{i+1}) such that

$$b \, T \, x_i \, T \, x_{i+1} \, T \, a,$$

we have $x_i \, I \, x_{i+1}$. The next proposition can easily be derived from arguments similar to the above ones. For the reader's convenience, we recall a few classical definitions. A graph is *strongly connected* if for all pairs of distinct nodes a, b, there is a path from a to b and a path from b to a in the graph. A *strongly connected component* of a graph is a maximal set of nodes on which the induced subgraph is strongly connected. The *induced subgraph* on a subset of nodes is a graph on this subset; its arcs are those of the original graph whose endpoints both belong to the subset.

Theorem 4.13 *The graph (A, N, H) is strongly connected iff for all i, $x_i \, I \, x_{i+1}$, where x_{i+1} denotes the immediate successor of x_i in order T. In any case, the nodes of each maximal simple path of (A, I) form a strongly connected component of (A, N, H) and the set of strongly connected components is totally ordered by P in the sense that for all pairs of distinct connected components C, C' :*

$$\forall x \in C, \, \forall y \in C', x \, P \, y$$

or

$$\forall x \in C, \, \forall y \in C', y \, P \, x.$$

To illustrate this result on an example, have a look at figure 4.5. There are two connected components in the super synthetic graph which are $\{ \, 1, 2, 3, 4 \, \}$ and $\{ \, 5, 6, 7, 8, 9 \, \}$. In the matrix associated to this semiorder (figure 4.3), it can be observed that this bipartition of the nodes is determined by the pair $(4, 5) \in P$. At that place, the 1's "hit" the diagonal of the matrix.

4.11   More about paths of SSG

Let $h_P(x)$ denote the height of $x \in A$ w.r.t. relation P, i.e. the length of the maximal length path of (A, P) whose origin is x; let $h(P)$ denote the maximal value of $h_P(x)$, $x \in A$.

Theorem 4.14 *For all $a, b \in A$,*
(i) $(a, b) \in T \Longrightarrow h_P(a) \geq h_P(b)$;
(ii) $(a, b) \in N \Longrightarrow h_P(b) = h_P(a) - 1$;
(iii) $(b, a) \in H \Longrightarrow h_P(a) \leq h_P(b) \leq h_P(a) - 1$.

The proof is to be found in section 4.16.

The third assertion of this lemma suggests that there are two types of hollows: arcs of H linking nodes of equal height ($h_P(a) = h_P(b)$) and arcs of H linking nodes with height difference equal to 1. The latter arcs will be of particular importance in this section and we denote them by H^*.

Definition 4.7 $(a, b) \in H^*$ *iff* $(a, b) \in H$ *and* $h_P(a) = h_P(b) + 1$.

The graph $(A, N \cup H^*)$ is a subgraph of (A, N, H) and its paths are those which are useful for the determination of a minimal representation. In particular, we have the following property.

Theorem 4.15 *The relation* $N \cup H^*$ *is acyclic; its transitive closure* $M = (N \cup H^*)^{tc}$ *is hence a strict partial order (transitive and asymmetric relation).*

We now consider also the height of the nodes in the relation $N \cup H^*$ or equivalently in M; we denote by $h_M(x)$ the maximal length of the paths of $N \cup H^*$ starting from x. We prove that there is always a path in $N \cup H^*$ from a to b between two elements of the same height h_P . This path goes from the smaller element w.r.t. T to the greater one.

Theorem 4.16 *If* aTb *and* $h_P(a) = h_P(b)$, *then there exists a path of* $N \cup H^*$ *from* a *to* b *and hence* $h_M(a) > h_M(b)$. *All such paths have null balance.*

The proof is in section 4.16.

The results above allow for a graphical representation of (A, M) which is particularly readable. Figure 4.6 is such a representation for the example of figure 4.3.

The nodes are represented according to the levels of h_P displayed in increasing order along the vertical axis. On each level the nodes appear according to order T in increasing order from left to right. The noses (arcs of N) are represented by an arrow directed downwards and linking consecutive levels while arcs of H^* are represented by arrows pointing upwards and linking a level of h_P to the level just above. The value of $h_M(x)$ is written between parentheses aside the label x of each node.

Comparing with $(A, N \cup H)$, one observes that the lacking hollows are exactly $(9, 7)$, $(4, 3)$ and $(2, 1)$; all of them link extreme nodes on a level of h_P, but the converse is not true since there can be levels $\{ x : h_P(x) = i \}$ whose extreme points are not linked by a hollow ($(6, 5)$ in our example). Let us denote by a_i (resp. z_i) the first (resp. the last) node on level $h_P(x) = i$.
Let $d = \frac{1}{2} \max\limits_{0 \leq i \leq h(P)} [h_M(a_i) - h_M(z_i)]$.

It results immediately from theorem 4.16 that

$$d = \frac{1}{2} \max \{ h_M(x) - h_M(y), \; x, y \in A, \; h_P(x) = h_P(y) \}.$$

It is now possible to exhibit paths of $N \cup H^*$ whose value yields a minimal representation for any acceptable value of threshold q.

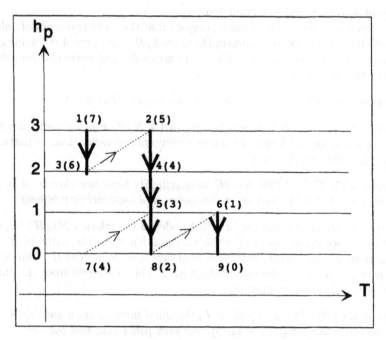

Figure 4.6: A representation of $(A, N \cup H^*)$

Theorem 4.17 *For any value of $q \geq d\varepsilon$, the function*

$$f(x) = h_P(x)(q + \varepsilon/2) + h_M(x).\varepsilon/2$$

is a (q, ε)-representation of S. Moreover, it is the minimal one and $d = \alpha$.

Note that $f(x)$ is the value of any path of length $h_M(x)$ in $(A, N \cup H^*)$, starting from x and ending in z_0 (the last node according to T). Indeed, the value $v(\Gamma)$ of such a path is

$$v(\Gamma) = |P \cap \Gamma|(q + \varepsilon) - |I \cap \Gamma|q.$$

Since $|I \cap \Gamma| = \frac{h_M(x) - h_P(x)}{2}$ and $|P \cap \Gamma| = \frac{h_M(x) + h_P(x)}{2}$, we have

$$v(\Gamma) = \frac{h_M(x) + h_P(x)}{2}(q + \varepsilon) + \frac{h_P(x) - h_M(x)}{2}q$$
$$= f(x).$$

The proof (in section 4.16) essentially consists in showing that f satisfies the systems of constraints 4.4 and 4.5.

Theorem 4.17 is not only a conceptually interesting result; it also has computational consequences. The computation of a minimal integer representation of a semiorder can be done in linear time and space (see Mitas 1994).

With the next results, we make a step further towards a precise description of the important paths of (A, S). The proofs of the next two theorems are given in section 4.16.

Theorem 4.18 *For all* $i = 1, ..., h(P)$, $z_i = \max_T \{ x : h_P(x) = i \}$ *is the origin of an arc* $(z_i, y) \in N$, *for some* y *such that* $h_P(y) = i - 1$.

Theorem 4.19 *Any path of length* $h_M(x)$ *starting from* x *in* $(A, N \cup H^*)$ *ends up in* z_0, *the last node w.r.t.* T. *The balance of such a path is* $h_P(x)$.

Theorem 4.20 *For all* $x \in A$, *for all* $q \geq \alpha\varepsilon$, *any path of maximal length among the paths starting from* x *in* $(A, N \cup H^*)$ *has maximal value among all paths in* $G(q, \varepsilon)$ *whose origin is* x.

The last result follows from theorems 4.17 and 4.19 which imply that $f(x)$ is the value of a path of maximal length of $(A, N \cup H^*)$, starting in x. This is a striking property since it allows to substitute a family of valued graphs by a single non-valued graph for computing numerical representations of semiorders.

This theorem gives full sense and generality to formulae such as those in the second row of table 4.2; these formulae yield the value of the minimal ε-representation for any threshold $q \geq \alpha\varepsilon$; they express the value of a path issued from each node and whose value remains maximal among all such paths whatever the value of the threshold (provided it is compatible with a representation, i.e. it is larger than $\alpha\varepsilon$).

Note also that a result similar to theorem 4.20 holds for the maximal valued paths finishing in any x of A. For the example of figure 4.3, we give the formulae for the maximal values of those paths in table 4.3; we denote those values by $g'_{q,\varepsilon}$.

	1	2	3	4	5	6	7	8	9
$g'_{q,\varepsilon}$	0	ε	$q + \varepsilon$	$q + 2\varepsilon$	$2q + 3\varepsilon$	$2q + 4\varepsilon$	$2q + 5\varepsilon$	$3q + 4\varepsilon$	$3q + 5\varepsilon$
$\overline{g}_{q,\varepsilon} - g_{q,\varepsilon}$	0	ε	$q + \varepsilon$	$q + 2\varepsilon$	$2q + 3\varepsilon$	$2q + 4\varepsilon$	$3q + 3\varepsilon$	$3q + 4\varepsilon$	$3q + 5\varepsilon$

Table 4.3: Maximal value of the paths finishing in each element of the semiorder of figure 4.3

It is readily seen that some formulae are found in both $g_{q,\epsilon}$ and $g'_{q,\epsilon}$ (not for the same $x \in A$) but not all. If we take the largest value $\overline{g}_{q,\epsilon}$ of $g_{q,\epsilon}$ over A (which is equal to the largest value of $g'_{q,\epsilon}$) and subtracting $g_{q,\epsilon}(x)$, for all $x \in A$, we get the third row of table 4.3; the third row is equal to the second one except for $x = 7$. Indeed there is a gap between $g_{q,\epsilon}(x)$ and $\overline{g}_{q,\epsilon} - g_{q,\epsilon}(x)$ for the elements of A which are not on (any) maximal length path of the graph $(A, N \cup H^*)$ and only for those elements. In our example, the only maximal length path through the graph goes through the elements $(1, 3, 2, 4, 5, 8, 6, 9)$, avoiding 7. For 7, the gap is $q - 2\epsilon$; it is null iff $q = \alpha = 2\epsilon$. There are examples of semiorders for which the gap for certain elements is strictly positive for all values of the threshold compatible with a representation.

4.12   Semiorders as interval orders and the representation of interval orders

A semiorder is in particular an interval order. Hence, it can be represented by intervals of various lengths while in the usual and most specific representation, all intervals have the same length which is the value of the threshold. The possibility of a representation with constant length intervals is a consequence of theorem 3.1 (vii) which, in the language of interval orders, says: "A semiorder can be represented by a family of intervals of the real line with the additional property that the ordering of the intervals left endpoints is the same as the ordering of the intervals right endpoints". This means that there is a third type of representation of a semiorder, by intervals with their left endpoints in the same order as their right endpoints, but with non-necessarily constant lengths.

More formally, to each element a of A, we associate an interval $(l(a), r(a))$ of the real line; this can be done by giving two functions

$$l : A \longrightarrow \mathcal{R}$$
$$r : A \longrightarrow \mathcal{R}$$

with, $\forall\, a \in A$,

(4.7) $l(a) \leq r(a)$.

Definition 4.8 *The pair (l, r) is a* general representation by intervals *of the interval order or the semiorder $S = P \cup I$ if, $\forall\, a, b \in A$,*

$$aPb \quad iff \quad l(a) > r(b).$$

For semiorders, we will also be interested in the more constrained representation which we call "specific" and is defined as follows.

Definition 4.9 *The pair (l, r) is a* specific representation by intervals *of the semiorder $S = P \cup I$ if, $\forall\, a, b \in A$,*

$$l(a) < l(b) \quad \Rightarrow \quad r(a) \leq r(b)$$

and

$$r(a) < r(b) \quad \Rightarrow \quad l(a) \leq l(b).$$

In contrast with the general and specific representations defined above, the usual representation of a semiorder will be referred to as "representation by intervals of constant length". These are a special case of "specific representation by intervals". In this section up to section 14, we will be concerned with the general and specific interval representations of a semiorder. The problem of finding a *general* interval representation of a semiorder S (or an interval representation of an interval order) is rather simple: it can be solved contructively in the following way.

General insertion procedure

Start for instance by assigning a position on the real line to all left endpoints of the intervals $(l(a), r(a))$ for $a \in A$. These points can be placed in any manner compatible with the weak order T_1 (see theorem 3.10) i.e.

$$aT_1b \quad \Rightarrow \quad l(a) > l(b).$$

In cases where $(a, b) \notin T_1 \cup T_1^{-1}$, we are free of setting $l(a) < l(b)$, $l(a) = l(b)$ or $l(a) > l(b)$; the second step consists in inserting the right endpoints $r(a)$ for all $a \in A$, at appropriate places given the positions of the left endpoints. For any $a \in A$, $r(a)$ can be set to any value in the interval $[\alpha, \beta[$, where

$$
\begin{aligned}
\alpha &= \max\{l(a), \max(l(x), x \text{ s.t. } aSx)\}, \\
\beta &= \min(l(x), x \text{ s.t. } xPa).
\end{aligned}
$$

Applying this procedure to the semiorder of figure 4.3 yields for instance the representation illustrated on figure 4.7.

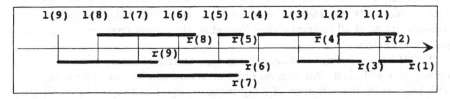

Figure 4.7: A general representation of the semiorder given in figure 4.3

Note that the ordering of the right endpoints need not be the same as the ordering of the left endpoints and that other representations of the same semiorder can be obtained by interchanging certain pairs of left or right interval endpoints. For instance in figure 4.7, the positions of $l(3)$ and $l(4)$ may be interchanged since neither $3T_14$ nor $4T_13$. The same can be done with any pair out of $r(5)$, $r(6)$ and $r(7)$.

The above procedure can easily be adapted for obtaining a *specific* interval representation for a semiorder.

Specific insertion procedure

Start by positionning the left endpoints just like in the general insertion procedure. When inserting the right endpoints, take into account the additional constraint that their ordering must be the same as the ordering of the left endpoints. More specifically, order the elements of A according to decreasing value of l (which is already fixed). The insertion starts with $l(a_1)$ maximal in $\{l(a), a \in A\}$; $r(a_1)$ can take any value not smaller than $l(a_1)$. Suppose that the first $(k-1)$ $r(a_j)$, $j < k$, have been inserted and turn to the insertion of $r(a_k)$. The right endpoint $r(a_k)$

can be set to any value in $[\alpha_k, \beta_k[$, where

$$\begin{aligned}
\alpha_k &= \max\{l(a_k), \max(l(x), \; x \text{ s.t. } a_k S x)\}, \\
\beta_k &= \min\{r(a_{k-1}), \min(l(x), \; x \text{ s.t. } x P a_k)\}.
\end{aligned}$$

In the rest of this chapter we will be concerned with the definition and the proof of existence of a *minimal specific* interval representation for a semiorder. In order to reach this goal, we have to study the interval representation of interval orders and prove the existence of minimal representations for interval orders.

4.13   Ordered structures associated with a step-type matrix

A fundamental observation about the insertions procedures described above is that the left and right endpoints of the intervals of $\mathcal{R}$ can be viewed as representing respectively the lines and columns of matrix M^S associated with S. In other words, the positions of these points only depend on the structure of the lines and columns of the matrix and not on their labels, i.e. on the elements of A they are associated with. This leads naturally to examine the connections between the (interval) representations of all interval orders (including the semiorders) associated with a particular matrix.

Let M be a $n \times n$ upper-diagonal step-type matrix of binary numbers denoted by m_{ij}, $i, j = 1, ..., n$. An interval order structure can be induced on any set A with $|A| = n$ by the definition of 2 bijections $\sigma_1, \sigma_2 : \{1, 2, ..., n\} \to A$. σ_1 (resp. σ_2) associates an element of A with each line (resp. column) of matrix M. The induced structure on A is defined by, $\forall a, b \in A$,

(4.8)
$$\begin{aligned}
aPb &\quad \text{iff} \quad m_{ij} = 1, \\
aIb &\quad \text{iff} \quad m_{ij} = m_{ji} = 0,
\end{aligned}$$

where $i = \sigma_1^{-1}(a)$ and $j = \sigma_2^{-1}(b)$.

In general, not any pair of bijections will lead to an interval order but if (P, I) is an interval order, in general, several pairs of bijections will define the same relation (P, I). These two points will be commented upon below. Note that in particular, when $\sigma_1 = \sigma_2$, (P, I) is a semiorder on A. Let us first illustrate these points on an example. Consider the matrix M in figure 4.8

	1	2	3	4
1	0	0	1	1
2	0	0	0	1
3	0	0	0	0
4	0	0	0	0

Figure 4.8: Example of an upper diagonal step-type matrix

Let $A = \{\, a, b, c, d \,\}$. Let us fix σ_1 as follows:

$$\sigma_1(1) = a, \; \sigma_1(2) = b, \; \sigma_1(3) = c, \; \sigma_1(4) = d.$$

This corresponds indeed to choosing a labelling for the elements of A. In principle there are 24 possibilities for σ_2 but it turns out that

• only 8 possibilities for σ_2 lead to an interval order since a may not be placed in columns 3 or 4 and b may not be placed in column 4, else one would have $a\,P\,a$ or $b\,P\,b$.

• these 8 possibilities correspond to only two distinct structures (up to a permutation of the labels of A), one interval order and one semiorder; exchanging the elements associated with columns 1 and 2 or with lines 3 and 4 (c and d) give isomorphic structures. Figures 4.9 and 4.10 represent the two induced structures. Tables 4.4 and 4.5 show the list of permutations σ_2 yielding these structures (up to a change in the labels).

In the above example we can see that permutations have to be excluded when they lead to $a\,P\,a$ for some $a \in A$. It turns out to be the only reason for excluding a permutation.

Theorem 4.21 *Consider the structure (P, I) defined on the set A by choosing two bijections*

$$\sigma_1, \sigma_2 : \{1, 2, \ldots, n\} \longrightarrow A$$

and using (4.8). The following conditions are equivalent:
(i) the structure (P, I) is an interval order;
(ii) $\forall a, b, \; a\,P\,b \Longrightarrow \sigma_1^{-1}(a) < \sigma_1^{-1}(b)$ and $\sigma_2^{-1}(a) < \sigma_2^{-1}(b)$;
(iii) $\forall a \in A, \; m_{\sigma_1^{-1}(a)\,\sigma_2^{-1}(a)} = 0.$

The proof of the theorem can be found in section 4.16.

Let us finally say one word about the set of permutations (σ_1, σ_2) defining the same interval order on A. If two lines i, i_{i+1} of matrix M are identical, two permutations σ_1, σ_1' which differ only by the fact that $\sigma_1(i) = \sigma_1'(i + 1)$ and $\sigma_1(i + 1) = \sigma_1'(i)$, coupled with any σ_2, will define the same structure on A. This can be extended to any number of identical lines of M and, *mutatis mutandis*, similar conclusions hold for σ_2 and identical columns.

4.14 Representations of all interval orders associated with a step-type matrix

All representations of an interval order associated with a given upper diagonal step-type matrix can be obtained from the representations of any other interval order associated with the same matrix.

Let $S = (P, I)$ and $S' = (P', I')$, a pair of interval orders on the set A obtained from matrix M by using respectively the bijections (σ_1, σ_2) and (σ_1', σ_2'). A representation (l, r) of S being given, we can derive a representation of S' by defining

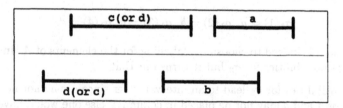

Figure 4.9: Semiorder induced by the permutations in table 4.4

	1	2	3	4
$\sigma_1 =$	a	b	c	d
$\sigma_2 =$	a	b	c	d
	a	b	d	c
	b	a	c	d
	b	a	d	c

Table 4.4: Permutations inducing the semiorder in figure 4.9

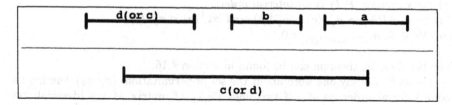

Figure 4.10: Interval order induced by the permutations in table 4.5

	1	2	3	4
$\sigma_1 =$	a	b	c	d
$\sigma_2 =$	a	c	b	d
	c	a	b	d
	a	d	b	c
	d	a	b	c

Table 4.5: Permutations inducing the interval order in figure 4.10

(l', r') as follows:

$$l'(a) = l(\sigma_1 \circ \sigma_1'^{-1}(a)),$$
$$r'(a) = r(\sigma_2 \circ \sigma_2'^{-1}(a)).$$

Since we know that S and S' are interval orders, we only need to verify that $aP'b$ iff $l'(a) > r'(b)$. This is readily done since we have

$$a\,P'\,b \quad \text{iff} \quad m_{\sigma_1'^{-1}(a)\sigma_2'^{-1}(b)} = 1$$
$$\text{iff} \quad \sigma_1 \circ \sigma_1'^{-1}(a)\ P\ \sigma_2 \circ \sigma_2'^{-1}(b)$$
$$\text{iff} \quad l(\sigma_1 \circ \sigma_1'^{-1}(a)) > r(\sigma_2 \circ \sigma_2'^{-1}(b)).$$

The claim follows immediately from the definition of l', r'.

Example

Consider the matrix associated with the semiorder on $A = \{1, 2, \ldots, 9\}$ defined in figure 4.3. The given semiorder $S = (P, I)$ corresponds to labelling the ith row (resp. j^{th} column), $i = 1, \ldots, 9$ (resp. $j = 1, \ldots, 9$) by "i" (resp. "j") i.e., the appropriate labelling of the rows and columns is obtained by taking $\sigma_1 = \sigma_2 =$ the identical permutation on A. Let $S' = (P', I')$ be the interval order defined on A by taking for σ_1', the identical permutation on A and σ_2' the permutation:

	1	2	3	4	5	6	7	8	9
σ_2'	2	1	4	3	7	5	6	9	8

Figure 4.11 a) shows a representation by intervals of the semiorder S. The chosen representation is derived from the minimal representation of S as a semiorder. More precisely, the left endpoints of the intervals are given by the minimal representation with $\varepsilon = 1$ and all intervals have the minimal (constant) length 2. Figure 4.11 b) shows a representation of S' which is derived from the representation of S through the changes described in the beginning of the present subsection. As σ_1' is the identical permutation, the left endpoints of the intervals associated to each element of A are common to both the representations of S and S'. The right endpoints have been exchanged according to σ_2', i.e. the right endpoint of "i" in the representation of S becomes the right endpoint of "$\sigma_2'(i)$" in the representation of S' (for instance $i = 5$ and $\sigma_2'(i) = 7$). Note that starting from any representation of S, the procedure yields a representation of S'.

4.15 Minimal representation of an interval order

The representation of S' induced by the minimal (constant threshold) representation of the semiorder S illustrated in figure 4.11 a) is not minimal in any sensible sense. Still, it is possible to define minimal (interval) representations of interval orders much in the same spirit as for semiorders. This was first observed in Doignon

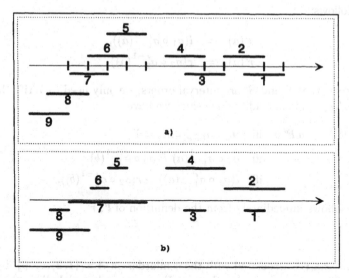

Figure 4.11: A semiorder and an interval order obtained through two different labellings of the lines and columns of the matrix in figure 4.3

1987. For another optimization problem about interval orders (minimal number of different interval lengths), see Fishburn 1985. Let us start with an alternative definition of interval order (valid for finite A).

Definition 4.10 *A reflexive relation $S = P \cup I$ on a finite set A is an interval order iff there exist a pair of functions $l, r : A \to \mathcal{R}^+$ and a positive constant ε such that, $\forall a, b \in A$,*

$$a\,P\,b \iff l(a) \geq r(b) + \varepsilon \;,$$
$$a\,I\,b \iff l(a) \geq r(b) \text{ and } l(b) \geq r(a) \;.$$

A triple (l, r, ε) satisfying these conditions is called an ε-representation of the interval order $P \cup I$.

Definition 4.11 *The ε-representation (l^*, r^*, ε) of the interval order $P \cup I$ is minimal iff for any other ε-representation (l, r, ε) (with the same ε), we have, $\forall a \in A$,*

$$l^*(a) \leq l(a) \;,$$
$$r^*(a) \leq r(a) \;.$$

The next theorem is taken from Doignon 1987 (Proposition 4); its proof is given in section 4.16.

Theorem 4.22 *For any interval order $P \cup I$, there exists a minimal ε-representation (l^*, r^*, ε); the values of l^* and r^* are integral multiples of ε.*

An immediate consequence of theorem 4.22 is worth mentioning. The minimal ε-representation of an interval order with $\varepsilon = 1$ is a representation on the smallest possible interval of the set of integer numbers.

Example

The minimal interval representation of the interval order illustrated in figure 4.11 b) is drawn in figure 4.12.

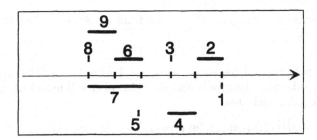

Figure 4.12: Minimal representation of the interval order in figure 4.11b)

4.16 Proofs of the theorems

Proof of theorem 4.2

From theorem 4.1, an ε-representation (g, q, ε) exists iff there is no circuit of positive value in $G(q, \varepsilon)$. The value of any circuit C in $G(q, \varepsilon)$ is

$$
\begin{aligned}
v(C) &= |C \cap P| (q + \varepsilon) + |C \cap I| (-q) \\
&= (|C \cap P| - |C \cap I|) q + |C \cap P| \varepsilon.
\end{aligned}
$$

As we know, by theorem 3.1 (vi), that

$$
|C \cap P| - |C \cap I| \leq -1,
$$

the fact that $v(C) \leq 0$ allows to derive the announced inequality.

Proof of theorem 4.4 (following Doignon 1988)

Lemma *Consider a sequence of d symbols "P" and d symbols "I" in such an order that, for any position, the number of "P" after the position is at least the number of "I" after the position. In such a case, for any integer e with $0 < e \leq d$, there is subsequence of 2e consecutive symbols which is composed of exactly e symbols "P" and e symbols "I".*

Proof of the lemma

As a direct consequence of the hypothesis, in the $2e$ first symbols, the balance of "P" and "I" is in favour of "I" and in the $2e$ last symbols, in favour of "P"; moreover we can suppose that the advantage is strict in both sequences, otherwise the lemma is proven. In any sequence of $2e$ consecutive symbols, the balance is an even number (positive or negative) and by shifting such a sequence, one unit left, the balance changes by $0, +2$ or -2 units. Hence, by a shift, the balance cannot jump from a strictly positive value (≥ 2) to a strictly negative value (≤ -2) without taking the value 0.

Proof of the theorem

Consider a maximal value path Ω starting from a. Suppose that Ω is not the announced Γ, i.e.
$$|\Omega \cap P| < |\Omega \cap I|.$$
Let Ω be the path $(a_1, a_2), (a_2, a_3), \ldots, (a_p, a_{p+1})$ with $a = a_1$. The arc (a_p, a_{p+1}) belongs to P (otherwise, remove it and the value of Ω will become larger). Let k be the largest index such that
$$|P \cap \{(a_k, a_{k+1}), (a_{k+1}, a_{k+2}), \ldots, (a_p, a_{p+1})\}| <$$
$$|I \cap \{(a_k, a_{k+1}), (a_{k+1}, a_{k+2}), \ldots, (a_p, a_{p+1})\}|.$$
We have $k < p-1$ and $(a_k, a_{k+1}) \in I$. Moreover
$$|P \cap \{(a_{k+1}, a_{k+2}), \ldots, (a_p, a_{p+1})\}| = |I \cap \{(a_{k+1}, a_{k+2}), \ldots, (a_p, a_{p+1})\}|.$$
Let d be the latter number. By hypothesis,
$$((a_k, a_{k+1}), (a_{k+1}, a_{k+2}), \ldots, (a_p, a_{p+1}))$$
has maximal value in the set of paths starting in a_k. To prove the theorem, we just need to prove the existence of a path H starting from a_k, with maximal value and such that
$$|H \cap P| \geq |H \cap I|$$
(if we can prove it, we repeat the procedure, getting smaller and smaller k, until there will be no "k" any more). Suppose that no such H does exist. This implies that the value of $((a_k, a_{k+1}), (a_{k+1}, a_{k+2}), \ldots, (a_p, a_{p+1}))$ which is
$$-q + d(q + \varepsilon) - dq = -q + d\varepsilon$$
is strictly positive. Hence, we have
$$d > \frac{q}{\varepsilon},$$
which means $d > \alpha$. Applying the above combinatorial lemma to the sequence of "P" and "I" of the path $((a_{k+1}, a_{k+2}), \ldots, (a_p, a_{p+1}))$, ensures that there is a subpath $((a_i, a_{i+1}), \ldots, (a_j, a_{j+1}))$ with $k+1 \leq i < j \leq p+1$ and
$$|P \cap \{(a_i, a_{i+1}), \ldots, (a_j, a_{j+1})\}| = |I \cap \{(a_i, a_{i+1}), \ldots, (a_j, a_{j+1})\}| = \lceil \alpha \rceil$$

with $\lceil \alpha \rceil$, the smallest integer strictly larger than α. It is clear that the arc (a_i, a_{j+1}) belongs to P, otherwise $(a_{j+1}, a_i) \in P \cup I$ and the existence of the circuit $\{a_i a_{i+1}, \ldots, a_j a_{j+1}, a_{j+1} a_i\}$ is incompatible either with the fact that S is a semiorder or with the definition of α. So, replacing $\{(a_i, a_{i+1}), \ldots, (a_j, a_{j+1})\}$ by (a_i, a_{j+1}) would not lessen the value of the path though it yields a path starting in a_k with at least as many arcs in P as in I.

Proof of theorem 4.5 (following Doignon 1988)

It is sufficient to show that α is an integer as, in view of theorem 4.3, g^* is characterized in terms of values of paths whose arcs weight either $-q^* = -\alpha\varepsilon$ or $q^* + \varepsilon = (\alpha + 1)\varepsilon$. By definition (see theorem 4.2), the value of α is the ratio

$$\alpha = \frac{|P \cap C|}{|I \cap C| - |P \cap C|}$$

for a certain circuit C. In the graph $G(\alpha\varepsilon, \varepsilon)$, the value $v(C)$ of circuit C is 0 (by definition of α). For any potential function of $G(\alpha\varepsilon, \varepsilon)$, the inequalities 4.1 associated with the arcs of C are equalities. Indeed, if $C = \{(a_0, a_1), (a_1, a_2), (a_2, a_3), \ldots, (a_{k-1}, a_k), (a_k, a_0)\}$, we have

$$
\begin{aligned}
g(a_0) \quad &\geq \quad g(a_1) \quad + \quad v(a_0, a_1) \\
&\geq \quad g(a_2) \quad + \quad v(a_1, a_2) \quad + \quad v(a_0, a_1) \\
&\vdots \quad\quad \vdots \quad\quad\quad \vdots \quad\quad\quad \vdots \quad\quad\quad \vdots \\
&\geq \quad g(a_k) \quad + \quad v(a_{k-1}, a_k) \quad + \quad \cdots \quad + \quad v(a_0, a_1) \\
&\vdots \quad\quad \vdots \quad\quad\quad \vdots \quad\quad\quad \vdots \\
&\geq \quad g(a_0) \quad + \quad v(C) .
\end{aligned}
$$

If any of these inequalities is strict, we get

$$g(a_0) > g(a_0) + v(C),$$

implying $v(C) < 0$ which is untrue.

There is no loss of generality in supposing that a_0 is the node where g assumes its minimal value on the set $\{a_0, \ldots, a_k\}$.

It is clear that $(a_0, a_1) \in I$, otherwise $g(a_0)$ would be larger than $g(a_1)$. Hence

$$g(a_1) = g(a_0) + \alpha\varepsilon .$$

Let f be the representation of the induced semiorder on $\{a_0, \ldots, a_k\}$ defined by

$$f(a_i) = g(a_i) - g(a_0), \quad \forall\, i = 0, \ldots, k .$$

We have $f(a_1) = \alpha\varepsilon$ which is the maximal value of the paths from a_1 in the subgraph on $\{a_0, \ldots, a_k\}$. By theorem 4.4 applied to the induced semiorder, we can find a maximal value path H of this graph, starting from a_1, with value $v(H) = \alpha\varepsilon$ and with

$$|P \cap H| \geq |I \cap H| .$$

Hence
$$v(H) = |P \cap H|(\alpha\varepsilon + \varepsilon) + |I \cap H|(-\alpha\varepsilon) = \alpha\varepsilon \ ,$$
which yields
$$\alpha = \frac{|P \cap H|}{|I \cap H| + 1 - |P \cap H|} \ .$$

Now, let y be the last node of H. Clearly $f(y) = 0$, $(y, a_1) \in I$ and $H \cup \{(y, a_1)\} = B$ is a circuit in $P \cup I$. Since B is a circuit, by theorem 3.1(vi),
$$|P \cap B| \le |I \cap B| - 1$$
and by construction of H,
$$|P \cap B| \ge |I \cap B| - 1.$$

Hence, $|I \cap B| - |P \cap B| = 1$ and $\alpha = |P \cap B|$ is an integer.

Proof of theorem 4.6

By definition of an ε-representation and by theorem 4.2, $q_{g^*} \le \alpha\varepsilon$ and $q'_{g^*} \ge \alpha\varepsilon + \varepsilon$. We prove that these inequalities are in fact equalities. In the graph $G(\alpha\varepsilon, \varepsilon)$ for which g^* is a potential function, the maximal value of a circuit is 0. Let $C = ((a_0, a_1), (a_1, a_2), (a_2, a_3), \ldots, (a_{k-1}, a_k)$, $(a_k, a_0))$ be such a circuit. The difference between the values of g^* at the extremities of any arc of C is exactly the value of the arc (see the proof of theorem 4.5, for further detail). Suppose, without loss of generality, that a_0 is the node of C where g^* is minimal. Hence, $(a_0, a_1) \in I$, $(a_k, a_0) \in P$ and
$$g^*(a_0) = g^*(a_1) - \alpha\varepsilon$$
and
$$g^*(a_k) = g^*(a_0) + \alpha\varepsilon + \varepsilon$$
which imply that $q_{g^*} = \alpha\varepsilon$, $q'_{g^*} \ge \alpha\varepsilon + \varepsilon$ and $\varepsilon_{g^*} = \varepsilon$.

To prove the second part of the theorem, we start with the formulation developed just after the statement of the theorem in section 4.8. Let $\tilde{g}^* = (K/\overline{g}^*) g^*$, $\tilde{\varepsilon} = (K/\overline{g}^*) 1$, $\tilde{q}^* = (K/\overline{g}^*) q^* = (K/\overline{g}^*) \alpha$, be the appropriate multiples of g^*, ε and q^* to be used in the $[0, K]$ interval. From the first part of the proof it is clear that $\tilde{q}^* = q_{\tilde{g}^*}$ and $\tilde{\varepsilon} = q'_{\tilde{g}^*} - q_{\tilde{g}^*}$. We have to prove that $\tilde{\varepsilon} \ge \varepsilon_g$, where ε_g is the contrast of any representation g of S in the $[0, K]$ interval. Let Γ be a maximal valued path in $G(\tilde{q}^*, \tilde{\varepsilon})$; we have
$$v(\Gamma) = \beta\tilde{q}^* + \gamma\tilde{\varepsilon}$$
for some non-negative integers β and γ; we also have
$$v(\Gamma) = K = \frac{K}{\overline{g}^*}[\beta\alpha + \gamma].$$

Obviously, the value of Γ in $G(q_g, \varepsilon_g)$ is $\beta q_g + \gamma\varepsilon_g \le K$ and, moreover, $q_g \ge \alpha\varepsilon_g$. From the above equalities and inequalities, we get
$$\tilde{\varepsilon} = \frac{K}{\overline{g}^*} \ge \frac{\beta q_g + \gamma\varepsilon_g}{\beta\alpha + \gamma} \ge \frac{\beta\alpha\varepsilon_g + \gamma\varepsilon_g}{\beta\alpha + \gamma} = \varepsilon_g.$$

Proof of theorem 4.7

a) Case of the "noses": $(a, b) \in N$.

We first prove that if (a, b) belongs to PPI, PIP or IPP, it doesn't belong to N. Suppose $(a, b) \in PPI$. Then there are c and d such that $aPcPdIb$. As $T = PI \cup IP$, we have aPc and cTb, with $c \neq b$, from which we get that $b \neq \min_T P^+(a)$ and $(a, b) \notin N$. The proof is similar in the cases $(a, b) \in PIP$ or IPP.

Suppose now that $(a, b) \notin N$ because $b \neq \min_T P^+(a)$. This implies that there exists $c \in A$, $c \neq b$, such that aPc and cTb. Hence we get that $(a, b) \in PIP \cup PPI$ since $T = PI \cup IP$. In a similar manner, starting from $a \neq \max P^-(b)$, we get the existence of $d \in A$, $d \neq a$, such that $aTdPb$ which means that $(a, b) \in PIP \cup IPP$.

b) Case of the "hollows": $(b, a) \in H$.

Suppose $(b, a) \in IIP$. There exist $c, d \in A$ such that $bIcIdPa$ which yields $c \in A \setminus P^-(b)$ and cTa. This implies $a \neq \max_T [A \setminus P^-(b)]$ and hence $(b, a) \notin H$. The same conclusion is reached when supposing $(b, a) \in IPI$ or PII. Conversely suppose $(b, a) \notin H$ and $a \neq \max_T [A \setminus P^-(b)]$. Then there is $c \in A$ such that cTa and $c \in A \setminus P^-(b)$. We have either bIc or bPc and hence (b, a) belongs to IPI, IIP, PPI or PIP. But $PPI \subseteq PI \subseteq IPI$ and $PIP \subseteq P \subseteq IPI$, which yields that $a \neq \max_T [A \setminus P^-(b)]$ and implies $(b, a) \in IIP \cup IPI \cup PII$. Similarly, suppose $b \neq \min_T [A \setminus P^+(a)]$. There is $d \in A$ such that bTd and $d \in A \setminus P^+(a)$. We have either dIa or dPa and hence, $(b, a) \in PII$, IPI, PIP or IPP. The same conclusion as before can be drawn.

Proof of theorem 4.8

Let Γ be a path with maximal balance among the paths from a to b in S and with maximal number of P-arcs among the set of paths with maximal balance from a to b in S. Suppose there is a $(P \setminus N)$-arc in Γ; due to theorem 4.7, this arc can be substituted by either PIP, $P^2 I$ or IP^2 without changing $\beta(\Gamma)$ yet increasing the number of P-arcs. Similarly if there were a $(I \setminus H)$-arc in Γ, one could substitute it by either $I^2 P$, PI^2 or IPI, without changing $\beta(\Gamma)$ yet increasing the number of P-arcs. This shows that Γ is a path of (A, N, H). Note also that a path which is not simple, i.e. visits some node more than once, cannot have maximal balance. Indeed, deleting the circuits leads to a strict increase of path balance (theorem 3.1(vi)).

Proof of theorem 4.9

It is immediate that aPb (resp. aTb) implies the existence of a path Γ from a to b in (A, N, H) with $\beta(\Gamma) > 0$ (resp. $\beta(\Gamma) \geq 0$). Indeed if aPb, the arc (a, b) is a path of (A, S) with balance equal to one. By theorem 4.8 we know that a path Γ from a to b in (A, S) which has maximal balance and the largest possible number of P arcs among the paths with maximal balance is a path of (A, N, H). The balance of such a path is greater or equal to one. A similar result holds when aTb is true since $T = PI \cup IP$ by definition (see section 3.16).

For proving the converse part of the theorem, suppose that Γ is a path from a to b in (A, N, H) with $\beta(\Gamma) > 0$ and that aPb is not true. This means that (b, a) belongs either to P or to I. Augmenting Γ with either the arc bPa or bIa yields a cycle Γ' with $\beta(\Gamma') \geq 0$ which contradicts theorem 3.1 (vi). Let us now prove that if Γ is a path from a to b in (A, N, H) with $\beta(\Gamma) \geq 0$, then aTb. In the valued graph associated with (A, S), the value of Γ which contains $|N \cap \Gamma|$ arcs of N and $|H \cap \Gamma|$ arcs of H is computed as

$$|N \cap \Gamma|(q + \varepsilon) - |H \cap \Gamma|q = \beta(\Gamma)q + |N \cap \Gamma|\varepsilon.$$

For any numerical representation g with appropriate value of the threshold q, the difference $g(a) - g(b)$ must be larger or equal to the value of any path from a to b. In the present case, $\beta(\Gamma) \geq 0$ implies that $g(a) \geq g(b)$ and hence aTb. Finally, the characterization of I results immediately from the characterization of P and from the completeness of S.

Proof of theorem 4.10

The "if" part is a direct consequence of theorem 4.9. Suppose $(a, b) \in I$. Due to the connexity of (A, N, H) (a mentioned consequence of theorem 4.8), there is at least one path from a to b and from b to a in (A, N, H). According to theorem 4.9(i) the balance of such a path is not positive otherwise one would have aPb or bPa. As (a, b) or (b, a) belongs to T and hence to $PI \cup IP$, there is at least one path of null balance from a to b or from b to a in (A, S); from theorem 4.8, the maximal balance path from a to b or from b to a (with maximal number of P-arcs) is in (A, N, H) and its balance is 0 (otherwise one would have aPb or bPa).

For proving the rest of the theorem suppose that there are both a path Γ from a to b and a path Γ' from b to a with $\beta(\Gamma) = \beta(\Gamma') = 0$. This means that there is a cycle of null balance passing by a and b which is possible only if $a = b$ and the cycle has no arc.

Proof of theorem 4.11

Let Γ be a path of maximal value from a to b in $G(q, \varepsilon)$. Suppose there is an arc of P which is not an arc of N in Γ. From theorem 4.7, we know that we can substitute it by a path with two P-arcs and one I-arc. This would result in augmenting the value of the path Γ by $2(q + \varepsilon) - q - (q + \varepsilon) = \varepsilon$, a contradiction. Similarly, if we suppose there is an arc of $I \setminus H$ in Γ, we can substitute it by a path with two I-arcs and one P-arc yielding an augmented value of Γ by $-2q + (q + \varepsilon) - (-q) = \varepsilon$, a contradiction. Hence Γ is made only of arcs of N and H.

Proof of theorem 4.14

(i) $(a, b) \in T$ implies $P^+(a) \supseteq P^+(b)$ and hence $h_p(a) \geq h_p(b)$.

(ii) As $(a, b) \in N$ implies $(a, b) \in P$, $h_p(a) \geq h_p(b) + 1$. Let c be the immediate successor of a on a maximal length P-path starting from a. Obviously, aTc and $h_p(a) = h_p(c) + 1$; c belongs to $P^+(a)$ and b is minimal in $P^+(a)$ w.r.t. T. Therefore, we have $aTbTc$ and hence $h_p(a) \geq h_p(b) \geq h_p(c)$. Finally from $h_p(c) = h_p(a) - 1 \geq h_p(b) \geq h_p(c)$, we get $h_p(b) = h_p(a) - 1$.

(iii) $(b, a) \in H$ implies aTb and hence $h_p(a) \geq h_p(b)$. Let c be the immediate successor of a on a maximal length P-path starting from a. Obviously $(a, c) \in T$ and $h_p(a) = h_p(c) + 1$. As $b = \max_T \{A \setminus P^+(a)\}$, we have $aTbTc$ and $h_p(a) \geq h_p(b) \geq h_p(c) = h_p(a) - 1$.

Proof of theorem 4.15

It is well-known that a relation is acyclic if and only if its transitive closure is an order i.e. is transitive and asymmetric. We prove that $N \cup H^*$ is acyclic by proving that the existence of a path of $N \cup H^*$ linking a to b where $h_p(a) = h_p(b)$ implies aTb. Acyclicity of $N \cup H^*$ immediately results from the irreflexivity of T by applying this property with $a = b$.

Suppose there is a path of $N \cup H^*$ from a to b where $h_p(a) = h_p(b)$. This path has an even number of arcs and as many from N as from H^* (as any arc of N goes one level down w.r.t. h_p while any arc of H^* goes one level up). From theorem 4.9(ii) we get immediately aTb.

Proof of theorem 4.16

Consider two consecutive elements x_i, x_{i+1} w.r.t. T with $h_p(x_i) = h_p(x_{i+1})$. If we can prove that for any such pair, $(x_i, x_{i+1}) \in M$, the theorem results.

As no two elements of A have the same predecessor and the same successor sets, let us suppose that $P^+(x_i) \neq^{\supset} P^+(x_{i+1})$. Let

$$a = \min_T \{P^+(x_i) \setminus P^+(x_{i+1})\} \text{ and } b = \max_T \{P^+(x_i) \setminus P^+(x_{i+1})\}.$$

Such a configuration is illustrated in figure 4.13.

Clearly, $(x_i, a) \in N$, $(b, x_{i+1}) \in H$ and even $(b, x_{i+1}) \in H^*$. Indeed $h_p(b) < h_p(x_i) = h_p(x_{i+1})$. If $a = b$, we have got the desired path $x_i N a H^* x_{i+1}$. Otherwise $a \neq b$ but all elements between a and b have the same predecessor set. Hence the reasonings about the pair (x_i, x_{i+1}) hold for each pair of immediate successors between a and b (inclusively). This recursive process stops after a finite number of steps as it always leads to elements of A one level below the preceding ones (w.r.t. h_p) and the number of such levels is finite. This finishes to prove the existence of a path from x_i to x_{i+1}. Obviously all paths linking two nodes of the same h_p height in $N \cup H^*$ have null balance.

Proof of theorem 4.17

In view of theorem 4.12 and in order to establish that f is a (q, ε) representation, it is sufficient to prove that it satisfies the systems of constraints (4.4) and (4.5).

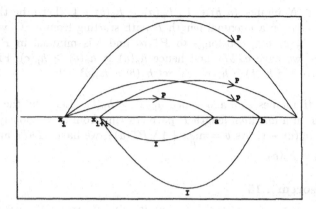

Figure 4.13: A configuration of the graph (A, P, I) of a semiorder

1) $(a, b) \in N \Rightarrow f(a) \geq f(b) + q + \varepsilon$.
If $(a, b) \in N$, $h_p(a) - h_p(b) = 1$ and $h_M(a) - h_M(b) \geq 1$.
Hence $f(a) - f(b) \geq q + \frac{\varepsilon}{2} + \frac{\varepsilon}{2} = q + \varepsilon$.

2) $(b, a) \in H \Rightarrow f(b) \geq f(a) - q$.

2.1) If $(b, a) \in H^*$, $h_p(b) - h_p(a) = -1$ and $h_M(b) - h_M(a) \geq 1$.
Hence $f(b) - f(a) \geq -(q + \frac{\varepsilon}{2}) + \frac{\varepsilon}{2} = -q$.

2.2) If $(b, a) \in H \backslash H^*$, $h_p(b) - h_p(a) = 0$ and $h_M(a) - h_M(b) \leq 2d$, due to the definition of d. Hence, $f(b) - f(a) \geq (-2d).\frac{\varepsilon}{2} = -d\varepsilon \geq -q$, by hypothesis.

Note that there is at least one level i on which

$$f(z_i) - f(a_i) = -d\varepsilon.$$

As $(z_i, a_i) \in I$, $q = d\varepsilon$ is the smallest possible value for q.

So we have established that f is a (q, ε) representation of S iff $q \geq d\varepsilon$. In the main text, after the theorem, we exhibit for all $x \in A$, a path whose value is exactly $f(x)$. This means that $f(x)$ is the minimal representation $g_{q,\varepsilon}$, $\forall q \geq d\varepsilon$. From theorem 4.2, we know that a necessary and sufficient condition for the existence of a (q, ε) representation is $q \geq \alpha\varepsilon$, for some integer constant α. From this we conclude that $d = \alpha$.

Proof of theorem 4.18

There exists y such that $h_p(y) = i - 1$ and $z_i P y$. Indeed there is a path from z_i in (A, P) which has i arcs of type P; let y be the first node on this path after z_i; it has height $h_P(y) = i - 1$. It is clear that $z_i = \max_{T}\{x : xPy\}$ since all x after z_i w.r.t.

T is such that $h_P(x) = i - 1$ and one may not have xPy with $h_P(x) = h_P(y)$. So z_i is the origin of a nose (z_i, y^*) with $y^* = \min\limits_T \{y : z_i Py\}$.

Proof of theorem 4.19

Supposing that a path of length $h_M(x)$ stops in some node $y \neq z_o$ leads to the conclusion that this path is not maximal in $(A, N \cup H^*)$, a contradiction. Consider two cases.

- If $h_P(y) \neq z_i$, $\forall i = 0, \ldots, h(P)$, by theorem 4.16, there is a path from y to z_i in $(A, N \cup H^*)$, hence the original path was not maximal in $(A, N \cup H^*)$.

- If $h_P(y) = z_i$ for some $i \neq 0$, theorem 4.19 shows that z_i is the origin of an arc of N and the original path ending in z_i was not maximal.

The second part of the theorem is a direct consequence of the first one since the balance of any path going from x to z_0 thrrough arcs of N or H^* is $h_P(x)$ due to the fact that using an arc of N (resp. of H^*) increases (resp. decreases) the balance by one unit.

Proof of theorem 4.21

$a) \Rightarrow c)$

By definition of P, if $\exists\, a$ such that $m_{\sigma_1^{-1}(a)\sigma_2^{-1}(a)} = 1$, then aPa.

$c) \Rightarrow b)$

Suppose aPb and $\sigma_1^{-1}(a) \geq \sigma_1^{-1}(b)$; aPb implies $m_{\sigma_1^{-1}(a)\sigma_2^{-1}(b)} = 1$ and , as M is a step-type matrix,

$$\forall i \leq \sigma_1^{-1}(a),\ m_{i\sigma_2^{-1}(b)} = 1 .$$

Hence $m_{\sigma_1^{-1}(b)\sigma_2^{-1}(b)} = 1$ which contradicts c). Similar conclusions can be drawn by supposing $\sigma_2^{-1}(a) \geq \sigma_2^{-1}(b)$.

$b) \Rightarrow a)$

Reconsider the general insertion procedure described in section 4.12. Let $l(a)$ be any function: $A \to \mathcal{R}$ compatible with the complete order induced on A by σ_1, i.e.

$$l(a) > l(b) \iff \sigma_1^{-1}(a) < \sigma_1^{-1}(b) .$$

Then we insert $r(a), \forall a$, as indicated in the procedure. By construction, we have $l(a) < r(a)$ and aPb iff $l(a) > r(b)$. Hence, the structure (P, I) is represented by the intervals $(l(a), r(a))$ in agreement with the definition of an interval order.

Proof of theorem 4.22

The conditions in definition 4.10 can be represented by means of a valued graph $(A, P \cup I)$ where the arcs of P receive the value ε and the arcs of I the value 0. As $P \cup I$ is an interval order, relation PI has no circuit. Hence it is possible to look at maximal value paths in the graph (A, PI) where the arcs (belonging to PI) have the value ε. Let us define

$$l^*(a) = \text{maximal value of the paths starting from } a \text{ in } (A, PI),$$
$$r^*(a) = \text{maximal value of the paths starting from } a \text{ with an}$$
$$I\text{-arc and going on with a path in } (A, PI).$$

Clearly l and r satisfy the conditions in definition 4.10 and are the smallest functions to do so.

Note that it was erroneously stated in Pirlot 1990 (Proposition 3.4), that the minimal difference between the evaluations of two distinct objects $\min\{|g(a) - g(b)|, \ a \neq b\}$ is maximal in the minimal representation. This statement is false as can be seen from the following example. Consider the semiorder on $A = \{a, b, c, d, e\}$ with $a \, P \, c$, $a \, P \, d$, $a \, P \, e$, $b \, P \, e$ and $c \, P \, e$. The minimal integer representation g^* with threshold $q^* = 2$ is

	a	b	c	d	e
g^*	6	4	3	2	0

The minimal value of the difference of two evaluations is $1 \ (= g^*(b) - g^*(c) = g^*(c) - g^*(d))$ while it is $1.5 - \eta$, for $0 < \eta < 0.25$, in the following representation g_η.

	a	b	c	d	e
g_η	6	$4-\eta$	3	$1+\eta$	0

VALUED SEMIORDERS

5.1 Introduction

Valued relations arise in fields like psychological studies on preference or discrimination, classification and decision-aid. Given a set A of elements $a, b, \ldots$, a value $v(a, b)$ may be associated to each oriented pair (a, b), representing either the proportion of times a given individual judges stimulus a to be "greater" than stimulus b or the proportion of individuals who prefer a to b, or else the "similarity" between a and b or the credibility or intensity of preference of a over b.

A very abundant literature is devoted to valued relations in mathematical psychology and in fuzzy sets theory (where "valued" is replaced by "fuzzy").

The purpose of this chapter is to study valued relations which can be considered as natural extensions of the semiorder structure. Some references on this subject are Doignon et al. 1986, Roberts 1971a, Roubens and Vincke 1985, Vincke 1978.

5.2 Definitions

Let A denote a finite set of elements $a, b, c, \ldots$ A *valued relation* v on the set A is a mapping from $A \times A$ into a bounded subset of $\mathcal{R}$; mapping into more general structures (see e.g. Monjardet 1984) may also be considered but we limit ourselves to numerical values. All the properties we will study here (with the exception of theorem 5.15) are invariant for a strictly monotone transformation so that it is not restrictive to consider the valued relation as a mapping from $A \times A$ into $[-1, +1]$.

In many applications, the knowledge of $v(a, b)$ implies that of $v(b, a)$, through a condition like

$$v(a, b) + v(b, a) = \text{ constant };$$

this is for example the case for the probabilistic relations π which are mappings from $A \times A$ to $[0, 1]$ satisfying

$$\pi(a, b) + \pi(b, a) = 1.$$

A valued relation will be said *ipsodual* if, $\forall\, a, b \in A$,

$$v(a, b) + v(b, a) = 0 = v(a, a).$$

Remark that an ipsodual valued relation can always be deduced from a probabilistic relation by putting

$$v(a, b) = 2\pi(a, b) - 1.$$

A λ-cut in a valued relation is the binary relation S^λ defined as follows:

$$aS^\lambda b \text{ iff } v(a, b) \geq \lambda.$$

Denoting by λ_i the values of v, ranking them in decreasing order $\lambda_1 > \lambda_2 > \ldots > \lambda_p$ and using them to define cuts yields the set of imbedded relations $\{S^{\lambda_i}, i = 1, 2, \ldots, p\}$ which is called the *chain of λ-cuts associated to v and is denoted* C^v. Remark that $S^{\lambda_p} = A \times A$.

If v is ipsodual, there is a bijection between the strictly positive and the non positive λ-cuts; for $\lambda_i > 0$,

$$
\begin{array}{rll}
aS^{\lambda_i}b & \text{iff} & v(a, b) \geq \lambda_i \\
& \text{iff} & v(b, a) \leq -\lambda_i \\
& \text{iff} & v(b, a) < -\lambda_{i+1} \\
& \text{iff} & v(b, a) \not\geq -\lambda_{i+1} \\
& \text{iff} & b\left(\neg S^{-\lambda_{i+1}}\right)a \\
& \text{iff} & a\left(S^{-\lambda_{i+1}}\right)^d b.
\end{array}
$$

Every strictly positive λ-cut is the dual of a non-positive λ-cut as explicited below (where $p = 2r - 1$):

$$\lambda_1 > \cdots \lambda_{r-1} > \lambda_r = 0 = -\lambda_r > -\lambda_{r-1} > \cdots > -\lambda_2 > -\lambda_1$$
$$\emptyset \subset S^{\lambda_1} \subset \cdots \subset S^{\lambda_{r-1}} \subset S^0 \subset S^{-\lambda_{r-1}} \subset \cdots \subset S^{-\lambda_2} \subset S^{-\lambda_1} = A \times A.$$

We denote by C_+^v the half-chain $\{S^{\lambda_i}, \lambda_i > 0\}$.

5.3   Graph and matrix representations of a valued relation

A valued relation on a set A may be represented by a complete and symmetric directed graph whose nodes are elements of A and where each edge (a, b) has a value $v(a, b)$. If the valued relation is ipsodual, it is not necessary to draw the loops and the negative edges.

A valued relation on a set A can also be represented by a matrix M^v where a row and a column are associated to each element of A and where, $\forall a, b \in A$,

$$M_{ab}^v = v(a, b).$$

The matrix is said *step-type* if it is possible to associate its rows and columns to the elements of A in such a way that its elements are non increasing from right to left in each row and from up to down in each column.

5.4 Semiordered valued relation

Theorem 5.1 *Assuming $v(a,a) = k$, $\forall\, a \in A$, the following conditions are equivalent:*

(i) $\forall\, a, b, c, d \in A$,

    (i1) $\max\{v(a,b), v(c,d)\} \geq \min\{v(a,d), v(c,b)\}$,

    (i2) $\max\{v(a,b), v(b,c)\} \geq \min\{v(a,d), v(d,c)\}$;

(ii) *the chain C^v of λ-cuts associated to v is such that, $\forall\, \lambda_i \leq k$, S^{λ_i} is a semiorder and $\forall\, \lambda_i > k$, S^{λ_i} is a strict semiorder.*

The proof is given in section 5.23.

Definition 5.1 *A valued relation satisfying the conditions of theorem 5.1 is called a semiordered valued relation.*

Particular case If $v(a,b) \in \{0,1\}$ and $k = 0$, then, defining

$$\left\{ \begin{array}{lll} aPb & \Longleftrightarrow & v(a,b) = 1, \\ aIb & \Longleftrightarrow & v(a,b) = v(b,a) = 0, \end{array} \right.$$

condition (i) in theorem 5.1 may be rewritten as

$$PIP \subset P \text{ and } P^2 I \subset P;$$

$S^0 = P \cup I$ is a semiorder and $S^1 = P$ is a strict semiorder (which is the dual of S^0).

5.5 Numerical representation of a semiordered valued relation

Theorem 5.2 *Let v be a valued relation. The following statements are equivalent:*

(i) v *is a semiordered valued relation with $v(a,a) = k$;*

(ii) *there exist real valued functions $g_1, g_2, \ldots, g_p, q_1, q_2, \ldots, q_p$, defined on A, such that, $\forall\, a, b \in A$,*

$$\left\{ \begin{array}{l} [q_i < 0 \text{ and } (aS^{\lambda_i}b \Longleftrightarrow g_i(a) \geq g_i(b) + q_i(b))] \text{ for } \lambda_i \leq k, \\ [q_i > 0 \text{ and } (aS^{\lambda_i}b \Longleftrightarrow g_i(a) > g_i(b) + q_i(b))] \text{ for } \lambda_i > k, \\ g_i(a) \geq g_i(b) \Longleftrightarrow g_i(a) + q_i(a) \geq g_i(b) + q_i(b); \end{array} \right.$$

(iii) *there exist real valued functions $g_1, g_2, \ldots, g_p$ and nonnegative constants $q_1, q_2, \ldots, q_p$ such that, $\forall\, a, b \in A$,*

$$\left\{ \begin{array}{l} aS^{\lambda_i}b \Longleftrightarrow g_i(a) \geq g_i(b) - q_i \text{ for } \lambda_i \leq k, \\ aS^{\lambda_i}b \Longleftrightarrow g_i(a) > g_i(b) + q_i, \text{ for } \lambda_i > k; \end{array} \right.$$

(iv) there exist real valued functions $g_1, g_2, \ldots, g_p$ *and a nonnegative constant* q
such that, $\forall\, a, b \in A$,

$$\begin{cases} aS^{\lambda_i}b \iff g_i(a) \geq g_i(b) - q \text{ for } \lambda_i \leq k, \\ aS^{\lambda_i}b \iff g_i(a) > g_i(b) + q \text{ for } \lambda_i > k. \end{cases}$$

This theorem is an immediate consequence of theorem 5.1 and of the results about numerical representations of semiorders and strict semiorders. The thresholds q_i being arbitrary, they can be all the same, leading to the last numerical representation.

Definition 5.2 *A* $(p+1)$-*tuple* $(g_1, g_2, \ldots, g_p, q)$ *verifying condition* (iv) *of theorem 5.2 is called a numerical representation constant threshold* q *of the semiordered valued relation* v.

Starting from a family of crisp semiorders (i.e binary relations which are semiorders), we can conversely state conditions under which they are the λ-cuts of a semiordered valued relation.

Theorem 5.3 *Let* $(S_i, i = 1, 2, \ldots, p)$ *be a strictly increasing chain of semiorders or strict semiorders, i.e. an indexed family of semiorders or strict semiorders on* A *such that* $\forall\, i = 1, 2, \ldots, p-1$,

$$S_i \subset S_{i+1};$$

suppose in addition that $S_p = A \times A$. *There exists a semiordered valued relation* v *such that* $(S_i, i = 1, 2, \ldots, p)$ *is the chain* C^v *of* λ-*cuts associated with* v.

The proof of this theorem is given in section 5.23. Note that the arguments used to prove theorem 5.3 also apply to prove that any chain of semiorders or strict semiorders, ordered by inclusion, can be viewed as a subset of all λ-cuts of a valued relation.

5.6   Example of a semiordered valued relation

v	a	b	c	d	e	f
a	0.5	0.9	0.5	0.1	0.8	1
b	0.4	0.5	0.3	0.2	0.8	0.7
c	0.4	0.5	0.5	0.1	0.8	0.7
d	0.5	0.9	0.6	0.5	0.8	0.7
e	0.3	0.5	0.2	0	0.5	0.5
f	0.3	0	0.1	0	0.8	0.5

The following tableaus present the chain of strict semiorders and of semiorders associated to v (except for $\lambda = 0$).

1-cut	a	b	c	d	e	f
a	0	0	0	0	0	1
b	0	0	0	0	0	0
c	0	0	0	0	0	0
d	0	0	0	0	0	0
e	0	0	0	0	0	0
f	0	0	0	0	0	0

0.9-cut	a	d	c	e	f	b
a	0	0	0	0	1	1
d	0	0	0	0	0	1
c	0	0	0	0	0	0
e	0	0	0	0	0	0
f	0	0	0	0	0	0
b	0	0	0	0	0	0

0.8-cut	a	d	c	f	b	e
a	0	0	0	1	1	1
d	0	0	0	0	1	1
c	0	0	0	0	0	1
f	0	0	0	0	0	1
b	0	0	0	0	0	1
e	0	0	0	0	0	0

0.7-cut	a	d	c	b	f	e
a	0	0	0	1	1	1
d	0	0	0	1	1	1
c	0	0	0	0	1	1
b	0	0	0	0	1	1
f	0	0	0	0	0	1
e	0	0	0	0	0	0

0.6-cut	d	a	c	b	f	e
d	0	0	1	1	1	1
a	0	0	0	1	1	1
c	0	0	0	0	1	1
b	0	0	0	0	1	1
f	0	0	0	0	0	1
e	0	0	0	0	0	0

0.5-cut	d	a	c	b	e	f
d	1	1	1	1	1	1
a	0	1	1	1	1	1
c	0	0	1	1	1	1
b	0	0	0	1	1	1
e	0	0	0	1	1	1
f	0	0	0	0	1	1

0.4-cut	d	c	a	b	e	f
d	1	1	1	1	1	1
c	0	1	1	1	1	1
a	0	1	1	1	1	1
b	0	0	1	1	1	1
e	0	0	0	1	1	1
f	0	0	0	0	1	1

0.3-cut	d	c	b	a	e	f
d	1	1	1	1	1	1
c	0	1	1	1	1	1
b	0	1	1	1	1	1
a	0	1	1	1	1	1
e	0	0	1	1	1	1
f	0	0	0	1	1	1

0.2-cut	d	b	c	a	e	f
d	1	1	1	1	1	1
b	1	1	1	1	1	1
c	0	1	1	1	1	1
a	0	1	1	1	1	1
e	0	1	1	1	1	1
f	0	0	0	1	1	1

0.1-cut	d	b	c	a	e	f
d	1	1	1	1	1	1
b	1	1	1	1	1	1
c	1	1	1	1	1	1
a	1	1	1	1	1	1
e	0	1	1	1	1	1
f	0	0	1	1	1	1

5.7   Minimal representation of a semiordered valued relation

From theorem 5.2 it is clear that v is a semiordered valued relation with $v(a, a) = k$ iff there exist real valued functions $g_1, g_2, \ldots, g_p$, a non-negative constant q and a positive constant ϵ such that, $\forall\, a, b \in A$,

$$
\begin{cases}
\text{for } \lambda_i \le k, \left[aS^{\lambda_i}b \Longrightarrow g_i(a) \ge g_i(b) - q\right] \text{ and} \\
\left[a\neg S^{\lambda_i}b \Longrightarrow g_i(a) \le g_i(b) - q - \epsilon\right]; \\
\text{for } \lambda_i > k, \left[aS^{\lambda_i}b \Longrightarrow g_i(a) \ge g_i(b) + q + \epsilon\right] \text{ and} \\
\left[a\neg S^{\lambda_i}b \Longrightarrow g_i(a) \le g_i(b) + q\right].
\end{cases}
$$

The set $\{g_1, g_2, \ldots, g_p, q, \epsilon\}$ is called an ϵ-representation of v. For the same reason as in section 4.2 we may fix ϵ.

Consider now the set $B = \bigcup_{i=1}^{p} A_i$ where $\forall\, i$, A_i is isomorph to A (we denote by a_i the image of a by this isomorphism) and $\forall\, i, j$, $A_i \cap A_j = \emptyset$; we build the valued graph $G(q, \epsilon)$ on B by giving

$$
\begin{cases}
\text{the value } q + \epsilon \text{ to the arcs } (a_i, b_i) \text{ such that } aS^{\lambda_i}b \text{ and } \lambda_i > k; \\
\text{the value } -q \text{ to the arcs } (b_i, a_i) \text{ such that } a\neg S^{\lambda_i}b \text{ and } \lambda_i > k; \\
\text{the value } -q \text{ to the arcs } (a_i, b_i) \text{ such that } aS^{\lambda_i}b \text{ and } \lambda_i \le k; \\
\text{the value } q + \epsilon \text{ to the arcs } (b_i, a_i) \text{ such that } a\neg S^{\lambda_i}b \text{ and } \lambda_i \le k.
\end{cases}
$$

Putting $g(a_i) = g_i(a)$, it is clear that $\{g_1, g_2, \ldots, g_p, q, \epsilon\}$ is an ϵ-representation of v iff g is a potential function of $G(q, \epsilon)$.

As the sets A_i are completely disconnected in the graph $G(q, \epsilon)$, it results from theorem 2.2 that there exists an ϵ-representation with threshold q iff

$$
\frac{q}{\epsilon} \ge \alpha = \max_{i} \alpha_i,
$$

where

$$
\alpha_i = \max_{C} \left\{ \frac{|C \cap P^i|}{|C \cap I^i| - |C \cap P^i|}, C \text{ circuit of } (A, S^{\lambda_i}) \right\} \text{ for } \lambda_i \le k,
$$

(P^i and I^i being respectively the asymmetric and the symmetric parts of S^{λ_i}) and

$$
\alpha_i = \max_{C} \left\{ \frac{|C \cap S^{\lambda_i}|}{|C \cap J^i| - |C \cap S^{\lambda_i}|}, C \text{ circuit of } (A, S^{\lambda_i} \cup J^i) \right\} \text{ for } \lambda_i > k,
$$

(J^i being the symmetric part of the dual of S^{λ_i}).

As a consequence, one can obtain a numerical representation of a semiordered valued relation through the following procedure:

1. choose ϵ;

2. choose a large enough value of the ratio $\frac{q}{\epsilon}$, for example, by taking $q \geq \max\{|S^{\lambda_1}|, |P^k|\}$;

3. for each a_i in A_i, for all i, find the maximal value $g(a_i)$ of the paths issued from a_i in the graph $G(q, \epsilon)$ (it can be solved separately in the restriction of $G(q, \epsilon)$ to each A_i);

4. put $g_i(a) = g(a_i), \forall\, a, \forall\, i$.

Definition 5.3 $\{g_1^*, g_2^*, \ldots, g_p^*, q, \epsilon\}$ *is minimal in the set of all non-negative ϵ-representations of a semiordered valued relation iff, $\forall\, a \in A$, for all non-negative ϵ-representation $\{g_1, g_2, \ldots, g_p, q, \epsilon\}$,*

$$g_i^*(a) \leq g_i(a).$$

Let $q^* = \alpha\epsilon$ with α defined as before. The next immediate result characterizes the minimal representation of a semiordered valued relation.

Theorem 5.4 *Applying the previous procedure with $q^* = q$ yields the minimal representation of the given semiordered valued relation.*

Conclusion It is clear that a minimal representation can be obtained separately for each λ-cut, leading to a numerical representation with different thresholds. If we want to use a unique threshold, we obtain a minimal representation of v taking, for this threshold, the largest value among the thresholds of the minimal representations of the λ-cuts.

5.8 Ipsodual semiordered valued relation

As an immediate consequence of theorem 5.1 and of the definition of C_+^v in the case of ipsoduality, we have the following result.

Theorem 5.5 *Assuming v is ipsodual, the following conditions are equivalent:*
(i) $\forall\, a, b, c, d \in A$,
$$\max\{v(a, b), v(c, d)\} \geq \min\{v(a, d), v(c, b)\},$$
$$\max\{v(a, b), v(b, c)\} \geq \min\{v(a, d), v(d, c)\};$$
(ii) C_+^v is a chain of strict semiorders.

5.9 Numerical representation of an ipsodual semiordered valued relation

The following theorem is an immediate corollary of theorem 5.2 and of ipsoduality.

Theorem 5.6 *Let v be an ipsodual valued relation. The following statements are equivalent:*

(i) v is an ipsodual semiordered valued relation;
(ii) there exist functions $g_1, g_2, \ldots, g_{r-1}, q_1, q_2, \ldots, q_{r-1}$, defined on A, such that, $\forall\, a, b \in A, \forall\, \lambda_i > 0, (i = 1, 2, \ldots, r - 1)$,

$$\begin{cases} aS^{\lambda_i}b \iff g_i(a) > g_i(b) + q_i(b), \\ q_i \geq 0, \\ g_i(a) \geq g_i(b) \implies g_i(a) + q_i(a) \geq g_i(b) + q_i(b); \end{cases}$$

(iii) there exist real valued functions $g_1, g_2, \ldots, g_{r-1}$ and nonnegative constants q_i such that, $\forall\, a, b \in A, \forall \lambda_i > 0, (i = 1, 2, \ldots, r - 1)$,

$$aS^{\lambda_i}b \iff g_i(a) > g_i(b) + q_i;$$

(iv) there exist real valued functions $g_1, g_2, \ldots, g_{r-1}$ and a nonnegative constant q such that, $\forall\, a, b \in A, \forall \lambda_i > 0, (i = 1, 2, \ldots, r - 1)$,

$$aS^{\lambda_i}b \iff g_i(a) > g_i(b) + q.$$

Note that for ipsodual semiordered valued relations, the semiorders $S^{-\lambda_i}$, $(i = 2, \ldots, r)$, can be represented by means of the function g_{i-1} and the opposite of the threshold used for representing their dual $S^{\lambda_{i-1}}$. So, for example, we could reformulate statement *(ii)* of theorem 5.6 as

(ii) there exist functions $g_1, g_2, \ldots, g_{r-1}, q_1, q_2, \ldots, q_{r-1}$, defined on A, such that, $\forall\, a, b \in A, \forall \lambda_i > 0, (i = 1, 2, \ldots, r - 1)$,

$$\begin{cases} aS^{\lambda_i}b \iff g_i(a) > g_i(b) + q_i(b), \\ aS^{-\lambda_{i+1}}b \iff g_i(a) \geq g_i(b) - q_i(b), \\ aS^{-\lambda_1}b, \forall\, a, b \in A, \\ q_i \geq 0, \\ g_i(a) \geq g_i(b) \implies g_i(a) + q_i(a) \geq g_i(b) + q_i(b). \end{cases}$$

In other words, if the vectors of functions $(g_1, g_2, \ldots, g_{r-1})$, $(q_1, q_2, \ldots, q_{r-1})$ satisfy condition *(ii)* of theorem 5.6, then the vectors of functions $(g_1, g_2, \ldots, g_{r-1}, g_{r-1}, \ldots, g_2, g_1, 0)$ and $(q_1, q_2, \ldots, q_{r-1}, -q_{r-1}, \ldots, -q_2, -q_1, 0)$, where 0 is the function which maps all elements of A onto 0, satisfy condition *(ii)* of theorem 5.2. Similar statements can be made about conditions *(iii)* and *(iv)* of these theorems.

The following result corresponds to theorem 5.3 in the case of ipsoduality.

Theorem 5.7 *Let $(S_i, i = 1, 2, \ldots, p)$ be a strictly increasing chain of strict semiorders (for $i \leq r, 0 \leq r \leq p$) or semiorders (for $i > r$), with $S_p = A \times A$. There exists an ipsodual semiordered valued relation v such that $(S_i, i = 1, 2, \ldots, p)$ is the chain $\mathcal{C}^v$ of λ-cuts associated with v iff $p = 2r - 1$ and $S_{r-i} = S^d_{r+i-1}$, for all $i = 1, \ldots, r - 1$.*

5.10   Example of an ipsodual semiordered valued relation

v	a	b	c	d	e	f
a	0	0.9	0.5	-0.5	0.8	1
b	-0.9	0	-0.5	-0.9	0.8	0.7
c	-0.5	0.5	0	-0.6	0.8	0.7
d	0.5	0.9	0.6	0	0.8	0.7
e	-0.8	-0.8	-0.8	-0.8	0	-0.8
f	-1	-0.7	-0.7	-0.7	0.8	0

The cuts at levels > 0.5 yield the same chain of strict semiorders as in the example of section 5.6.

5.11   Minimal representation of an ipsodual semiordered valued relation

Applying to an ipsodual semiordered valued relation, the procedure described in section 5.7 for building a minimal ϵ-representation of any semiordered valued relation yields an ϵ-representation of the type just described by the end of the preceding section. This means that the minimal ϵ-representation $(g_1^*, g_2^*, \ldots, g_{r-1}^*, g_r^*, \ldots, g_p^*; q^*, \epsilon)$ is in fact of the following particular form $(g_1^*, g_2^*, \ldots, g_{r-1}^*, g_{r-1}^*, \ldots, g_2^*, g_1^*, 0; q^*, \epsilon)$, with $p = 2r - 1$, i.e. $g_{r-1+i}^* = g_{r-i}^*$, for $i = 1, \ldots, r-1$, and $g_p^* = 0$.

5.12   Linear semiordered valued relation

We saw in section 5.4 that a semiordered valued relation is equivalent to a chain of semiorders and strict semiorders. An interesting question is to characterize the case where there is a common complete order associated to all these semiorders and strict semiorders. In other terms, we consider the situation where it is possible to associate the rows and columns of a matrix to the elements of A in such a way that all the matrices representing the semiorders and strict semiorders are step-type. If this is the case, the chain is called *homogeneous*. The following theorem is proved in section 5.23.

Theorem 5.8 *Assuming $v(a, a) = k$, $\forall a \in A$, the following conditions are equivalent:*
(i) $\forall\, a, b, c, d \in A$,

$$[v(c, a) < v(c, b) \text{ or } v(a, c) > v(b, c)]$$
$$\Rightarrow [v(d, a) \leq v(d, b) \text{ and } v(a, d) \geq v(b, d)];$$

(ii) the relation C defined by

$$aCb \text{ iff } \exists c : v(a, c) > v(b, c) \text{ or } v(c, a) < v(c, b)$$

is a weak order;

(iii) C^v is an homogeneous chain of semiorders (for $\lambda_i \leq k$) and of strict semiorders (for $\lambda_i > k$);

(iv) v can be represented by a step-type matrix.

Definition 5.4 *A valued relation satisfying the conditions of theorem 5.8 is called a linear semiordered valued relation.*

For completeness, we state the corresponding theorem for chains of semiorders.

Theorem 5.9 *Let $(S_i, i = 1, 2, \ldots, p)$ be a strictly increasing chain of semiorders or strict semiorders with $S_p = A \times A$. There exists a linear semiordered valued relation v such that $(S_i, i = 1, 2, \ldots, p)$ is the chain C^v of λ-cuts associated with v iff there is a common complete order associated with all S_i.*

5.13 Numerical representation of a linear semiordered valued relation

The following theorem is proved in section 5.23.

Theorem 5.10 *Let v be a valued relation. The following statements are equivalent:*

(i) v is a linear semiordered valued relation with $v(a, a) = k$;

(ii) there exist real valued functions $g, q_1, q_2, \ldots, q_p$, defined on A, such that, $\forall$ $a, b \in A$,

$$
\begin{cases}
q_i < 0 \text{ and } (aS^{\lambda_i}b \Longleftrightarrow g(a) \geq g(b) + q_i(b)), \text{ for } \lambda_i \leq k, \\
q_i > 0 \text{ and } (aS^{\lambda_i}b \Longleftrightarrow g(a) > g(b) + q_i(b)), \text{ for } \lambda_i > k, \\
g(a) \geq g(b) \Longleftrightarrow g(a) + q_i(a) \geq g(b) + q_i(b), \\
q_1(a) \geq q_2(a) \geq \cdots \geq q_p(a);
\end{cases}
$$

(iii) there exist real-valued functions $g_1, g_2, \ldots, g_p$ and a nonnegative constant q such that, $\forall$ $a, b \in A$,

$$
\begin{cases}
aS^{\lambda_i}b \Longleftrightarrow g_i(a) \geq g_i(b) - q \text{ for } \lambda_i \leq k, \\
aS^{\lambda_i}b \Longleftrightarrow g_i(a) > g_i(b) + q \text{ for } \lambda_i > k, \\
\forall i, j, \quad g_i(a) > g_i(b) \Rightarrow g_j(a) > g_j(b).
\end{cases}
$$

5.14 Example of a linear semiordered valued relation

v	a	b	c	d	e	f
a	0.5	0.6	0.8	0.9	0.9	1
b	0.5	0.5	0.7	0.7	0.8	0.8
c	0.4	0.4	0.5	0.6	0.7	0.8
d	0.3	0.4	0.5	0.5	0.7	0.8
e	0.1	0.3	0.4	0.5	0.5	0.8
f	0.1	0.2	0.2	0.3	0.4	0.5

Keeping the same complete order on the set $\{a, b, c, d, e, f\}$ each cut at a level > 0.5 clearly gives a step-type upper-diagonal matrix and each cut at a level ≤ 0.5 gives a lower-diagonal step-type matrix.

5.15 Minimal representation of a linear semiordered valued relation

Applying to a linear semiordered valued relation the procedure described in section 5.7, we obtain a minimal ϵ-representation which satisfies condition *(iii)* of theorem 5.10. The following property is proved in section 5.23.

Theorem 5.11 *The minimal ϵ-representation $(g_1^*, g_2^*, \ldots, g_p^*; q^*, \epsilon)$ of a linear semiordered valued relation satisfies*

$$\forall a, b \in A, \ \forall i, j = 1, \ldots, p, \quad [g_i^*(a) > g_i^*(b)] \Rightarrow [g_j^*(a) \geq g_j^*(b)] \ .$$

5.16 Ipsodual linear semiordered valued relation

The following theorems are corollaries of theorems 5.8 and 5.9 and of ipsoduality.

Theorem 5.12 *Assuming v is an ipsodual valued relation, the following conditions are equivalent:*
(i) $\forall \ a, b, c, d \in A$, $v(a, c) > v(b, c) \Rightarrow v(a, d) \geq v(b, d)$;
(ii) the relation C defined by

$$aCb \ \text{iff} \ \exists c : v(a, c) > v(b, c)$$

is a weak order;
(iii) C_+^v is an homogeneous chain of strict semiorders;
(iv) v can be represented by a step-type matrix.

Theorem 5.13 *Let $(S_i, i = 1, 2, \ldots, p)$ be a strictly increasing chain of strict semiorders (for $i \leq r, 0 \leq r \leq p$) or semiorders (for $i > r$), with $S_p = A \times A$. There exists an ipsodual linear semiordered valued relation v such that $(S_i, i = 1, 2, \ldots, p)$ is the chain C^v of λ-cuts associated with v iff $p = 2r - 1$, $S_{r-i} = S_{r+i-1}^d$, for all $i = 1, \ldots, r - 1$ and there is a common complete order associated with all S_i.*

5.17 Numerical representation of an ipsodual linear semiordered valued relation

The theorem below is a corollary of 5.10 and of ipsoduality.

Theorem 5.14 *Let v be an ipsodual valued relation. The following statements are equivalent:*
(i) v is an ipsodual linear semiordered valued relation;
(ii) there exist real-valued functions $g, q_1, q_2, \ldots, q_{r-1}$, defined on A, such that, $\forall$ $a, b \in A$, $\forall \lambda_i > 0$,

$$\begin{cases} aS^{\lambda_i}b \iff g(a) > g(b) + q_i(b), \\ q_i \geq 0, \forall i = 1, \ldots, r - 1, \\ g(a) \geq g(b) \iff g(a) + q_i(a) \geq g(b) + q_i(b), \\ q_1(a) \geq q_2(a) \geq \cdots \geq q_{r-1}(a); \end{cases}$$

(iii) there exist real-valued functions $g_1, g_2, \ldots, g_{r-1}$ and a nonnegative constant q such that, $\forall a, b \in A$, $\forall \lambda_i > 0$,

$$\begin{cases} aS^{\lambda_i}b \iff g_i(a) > g_i(b) + q, \\ \forall i, j, \quad g_i(a) > g_i(b) \Rightarrow g_j(a) \geq g_j(b). \end{cases}$$

5.18 Example of ipsodual linear semiordered valued relation

v	a	b	c	d	e	f
a	0	0.3	0.5	0.6	0.8	0.9
b	-0.3	0	0.4	0.4	0.5	0.7
c	-0.5	-0.4	0	0.4	0.5	0.6
d	-0.6	-0.4	-0.4	0	0.2	0.4
e	-0.8	-0.5	-0.5	-0.2	0	0.1
f	-0.9	-0.7	-0.6	-0.4	-0.1	0

Keeping the same complete order on the set $\{a, b, c, d, e, f\}$ each cut at level > 0 clearly gives a step-type upper-diagonal matrix.

5.19 Minimal representation of an ipsodual semi-ordered valued relation

The results of section 5.11 can easily be specialized to the present case taking into account theorem 5.11.

5.20 Chains of semiorders representable by a single function and constant thresholds

An especially important particular case of ipsodual linear semiordered valued relations arises when the values $v(a, b)$ are differences of numerical evaluations of a and b, i.e. when there is a function $g : A \longrightarrow R$ such that

$$v(a, b) = g(a) - g(b).$$

In this case, $v(a, b) > \lambda_i$ iff $g(a) > g(b) + \lambda_i$. This induces a numerical representation of all relations S^{λ_i} by means of a single function g and a series of thresholds $\lambda_1, \ldots, \lambda_p$.

Theorem 5.15 *Let v be a valued relation. The following statements are equivalent:*
(i) $\forall\, a, b, x \in A,\ v(a, b) = v(a, x) - v(b, x);$
(ii) there exist a real valued function $g : A \longrightarrow R$ such that
$v(a, b) = g(a) - g(b)$ *and nonnegative constants $q_1, \ldots, q_p$, such that*

$$\begin{cases} a S^{\lambda_i} b \Longleftrightarrow g(a) \geq g(b) - q_i, & \text{for } \lambda_i \leq 0, \\ a S^{\lambda_i} b \Longleftrightarrow g(a) > g(b) + q_i, & \text{for } \lambda_i > 0. \end{cases}$$

Much more interesting and difficult is the characterization of chains of semiorders that can be represented by a single function and constant thresholds. These structures have been investigated in the theory of probabilistic consistency (Roberts 1971a, Roberts 1979). In the particular case of two thresholds, they also arise in the context of preference modelling (Roy and Hugonnard 1982, Roy and Vincke 1987).

Definition 5.5 *A chain S_i, $i = 1, 2, \ldots, p$ of strict semiorders (for $i \leq r$, $0 \leq r \leq p$) and semiorders (for $i > r$) is representable by a single function and constant thresholds if there is a function $g : A \longrightarrow R$ and nonnegative numbers $q_1, \ldots, q_p$ such that, $\forall\, a, b \in A$,*

$$\begin{cases} \text{for } i \leq r,\, a S_i b \Longleftrightarrow g(a) > g(b) + q_i, \\ \text{for } i > r,\, a S_i b \Longleftrightarrow g(a) \geq g(b) - q_i. \end{cases}$$

Conditions under which a chain of semiorders is representable by constant thresholds were first investigated in Cozzens and Roberts 1982 for the two-threshold case. A characterization of the general case can be found in Doignon 1987; we follow the latter paper in the rest of this section.

Just as we did for a single semiorder in section 5.7, we define a notion of ϵ-representation (with constant thresholds).

Definition 5.6 *Let g be a real-valued function on A; $q_1,\ldots,q_p$, a set of non-negative numbers and ϵ, a positive number. Let S_i, $i = 1,2,\ldots,p$ a chain of strict semiorders (for $i \leq r$, $0 \leq r \leq p$) and semiorders (for $i > r$). The set $\{g, q_1, q_2, \ldots, q_p, \epsilon\}$ is called a single function ϵ-representation with constant thresholds $q_1, \ldots, q_p$, of the chain of semiorders if, $\forall\, a, b \in A$,*

$$\begin{cases} \text{for } i \leq r, & (aS_ib \Rightarrow g(a) \geq g(b) + q_i + \epsilon) \\ & \text{and } (a\neg S^{\lambda_i}b \Rightarrow g(a) \leq g(b) + q_i); \\ \text{for } i > r, & (aS_ib \Rightarrow g(a) \geq g(b) - q_i) \\ & \text{and } (a\neg S_ib \Rightarrow g(a) \leq g(b) - q_i - \epsilon). \end{cases}$$

Note that in the above definition, all thresholds are defined as non-negative; the opposite of a threshold is used when the threshold is associated with a (non-strict) semiorder. It is clear that for finite A, there exists an ϵ-representation of a chain of semiorders iff there exists a representation by a single function and constant thresholds.

Generalizing the graph $G(q, \epsilon)$ of section 5.7, we define a valued multiple graph $G(q_1, \ldots, q_p, \epsilon)$ where each inequality involved in definition 5.5 is represented by a valued arc; namely, $\forall a, b \in A$,

$$\begin{cases} g(a) \geq g(b) - q_i & \text{is represented by an arc } (a,b) \text{ whose value} \\ & \text{is } -q_i, \\ g(a) \geq g(b) + q_i + \epsilon & \text{is represented by an arc } (b,a) \text{ whose value} \\ & \text{is } q_i + \epsilon. \end{cases}$$

By theorem 4.1, $G(q_1, \ldots, q_p, \epsilon)$ admits a potential function iff there is no circuit of strictly positive value. Let P_i denote S_i for $i \leq r$ and the dual of S_i for $i > r$. Let P_i^d denote the dual of P_i. This means that all P_i, $i = 1, \ldots, p$, are strict semiorders and all P_i^d, $i = 1, \ldots, p$, are the corresponding (complete) semiorders. The non-valued version of $G(q_1, \ldots, q_p, \epsilon)$ will also be considered; $\mathcal{P}$ denotes the multiple graph on A whose set of arcs is $\bigcup_{i=1}^p (P_i \cup P_i^d)$.

In $G(q_1, \ldots, q_p, \epsilon)$, the value associated to a P_i-arc is $q_i + \epsilon$ while the value attached to a P_i^d-arc is $-q_i$. The graph admits a potential function iff the value of any circuit C is non-positive which can be written as follows: for all circuit C,

$$\sum_{i=1}^p |\, P_i \cap C\,|(q_i + \epsilon) + \sum_{i=1}^p |\, P_i^d \cap C\,|(-q_i) \leq 0$$

or

$$(5.1) \qquad \sum_{i=1}^p (|\, P_i^d \cap C\,| - |\, P_i \cap C\,|)\, q_i - \epsilon \sum_{i=1}^p |\, P_i \cap C\,| \geq 0.$$

For circuits without P_i-arcs, this condition is trivially fulfilled so that we can exclude such circuits from consideration. If $|\, P_i \cap C\,| \neq 0$, since $\epsilon > 0$, the latter inequality is tantamount

$$(5.2) \qquad \sum_{i=1}^p (|\, P_i^d \cap C\,| - |\, P_i \cap C\,|)\, q_i > 0.$$

We hence have shown the following theorem.

Theorem 5.16 *There exists an ϵ-representation of a chain C_i, $i = 1,\ldots,p$, of semiorders or strict semiorders with constant thresholds $q_1, q_2, \ldots, q_p$ iff for all non-empty circuit C of the graph $\mathcal{P}$, we have*

$$\sum_{i=1}^{p} \left(\mid P_i^d \cap C \mid - \mid P_i \cap C \mid \right) q_i > 0.$$

The latter theorem does not provide a characterization of chains of semiorders representable by some constant threshold vector but rather of chains of semiorders representable by a *given* vector of thresholds. In order to get a complete and testable characterization, the following definition is introduced.

Definition 5.7 *A k-cyclone is a non-empty union of at most k-circuits of the graph $\mathcal{P}$. A k-cyclone C is balanced iff $\forall i = 1, \ldots, p$,*

$$\mid C \cap P_i \mid = \mid C \cap P_i^d \mid.$$

Theorem 5.17 *There is a constant threshold representation of the chain C_i, $i = 1, \ldots, p$, of semiorders or strict semiorders iff no p-cyclone of the graph $\mathcal{P}$ is balanced.*

This theorem, which is proved in section 5.23, provides a way of testing in practice the existence of a representation with constant thresholds. This opportunity is exploited in the MACBETH method which aims at building a value function on a set of alternatives on the basis of pairwise comparisons of alternatives (cf. Bana e Costa and Vansnick 1994). Note that the proof of theorem 5.17 holds when only p-cyclones of noses and hollows of P_i are considered. This remark makes it easier to test for the existence of p-cyclones.

5.21 Examples of chains of semiorders representable (or not) by a single function and constant thresholds

We first consider the following family of three semiorders or strict semiorders $S_1 \subset S_2 \subset S_3$ on $A = \{a, b, c, d, e\}$

S_1	a	b	c	d	e
a	0	0	0	0	1
b	0	0	0	0	0
c	0	0	0	0	0
d	0	0	0	0	0
e	0	0	0	0	0

S_2	a	b	c	d	e
a	0	1	1	1	1
b	0	0	0	1	1
c	0	0	0	0	1
d	0	0	0	0	0
e	0	0	0	0	0

S_3	a	b	c	d	e
a	1	1	1	1	1
b	1	1	1	1	1
c	0	1	1	1	1
d	0	1	1	1	1
e	0	0	1	1	1

The chain (S_1, S_2, S_3) is representable by the function

	a	b	c	d	e
g	5	3	2	1	0

and the thresholds $q_1 = 4$, $q_2 = 1$, $q_3 = 2$ (the latter threshold being used to define a semiorder, not a strict semiorder) ; it is a subchain of the chain C^v associated with the ipsodual linear valued relation v defined by $v(x, y) = g(x) - g(y)$

v	a	b	c	d	e
a	0	2	3	4	5
b	-2	0	1	2	3
c	-3	-1	0	1	2
d	-4	-2	-1	0	1
e	-5	-3	-2	-1	0

S_1 (resp. S_2, S_3) corresponds to the λ-cut of v for $\lambda_1 = 5$ (resp. $\lambda_2 = 2$, $\lambda_3 = -2$).

An example of a chain which is not representable by a single function and constant thresholds is obtained by slightly modifying S_1 into $\widetilde{S}_1$

$\widetilde{S}_1$	a	b	c	d	e
a	0	0	0	0	1
b	0	0	0	0	1
c	0	0	0	0	0
d	0	0	0	0	0
e	0	0	0	0	0

The chain $(\widetilde{S}_1, S_2, S_3)$ is not representable by a single function because one can find a 1-cyclone which is balanced in $\mathcal{P}$; we draw in figure 5.1 the subgraph of $\mathcal{P}$ where only the noses and hollows of $\widetilde{S}_1$, S_2, $P_3 = S_3^d$ are represented. (In case x and y are equivalent with respect to some S_i, and x is involved in a nose or a hollow, it is necessary to draw the corresponding arc but also the arc where x is substituted by y.)

The 1-cyclone (which is a simple circuit) a, b, e, d, a with $(a, b) \in P_2$, $(b, e) \in P_1$, $(e, d) \in P_2^d$ and $(d, a) \in P_1^d$ is balanced. The corresponding constraints in $G(q_1, q_2, q_3, \epsilon)$ are

$$
\begin{aligned}
g(a) &\geq g(b) + q_2 + \epsilon, \\
g(b) &\geq g(e) + q_1 + \epsilon, \\
g(e) &\geq g(d) - q_2, \\
g(d) &\geq g(a) - q_1.
\end{aligned}
$$

Summing up these constraints, one gets $0 \geq 2\epsilon$ which can not be satisfied for any $\epsilon > 0$.

5.22   Minimal representation

An interesting question which is left unanswered is the existence of a *minimal* ϵ-representation for chains of semiorders representable by a single function and

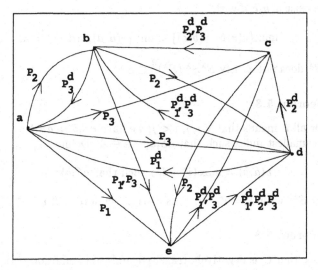

Figure 5.1: Graph of noses and hollows

constant thresholds.

5.23 Proofs of the theorems

Proof of theorem 5.1

$(i) \Rightarrow (ii)$

- $\forall \lambda_i,\ \forall a,b,c,d,\ aS^{\lambda_i}b,\ b\left(S^{\lambda_i}\right)^d c,\ cS^{\lambda_i}d$ imply $v(a,b) \geq \lambda_i$, $v(c,b) < \lambda_i$, $v(c,d) \geq \lambda_i$, and by property $(i1)$,
  $\max\{v(a,d), v(c,b)\} \geq \min\{v(a,b), v(c,d)\} \geq \lambda_i$,
  hence $v(a,d) \geq \lambda_i$ or $aS^{\lambda_i}d$, so that $S^{\lambda_i}\left(S^{\lambda_i}\right)^d S^{\lambda_i} \subset S^{\lambda_i}$.

- $\forall \lambda_i,\ \forall a,b,c,d,\ aS^{\lambda_i}b,\ bS^{\lambda_i}c,\ c\left(S^{\lambda_i}\right)^d d$, imply $v(a,b) \geq \lambda_i$, $v(b,c) \geq \lambda_i$, $v(d,c) < \lambda_i$, and by property $(i2)$,
  $\max\{v(a,d), v(d,c)\} \geq \min\{v(a,b), v(b,c)\} \geq \lambda_i$,
  hence $v(a,d) \geq \lambda_i$ or $aS^{\lambda_i}d$, so that $S^{\lambda_i}S^{\lambda_i}\left(S^{\lambda_i}\right)^d \subset S^{\lambda_i}$.

- Moreover, it is clear that S^{λ_i} is reflexive $\forall \lambda_i \leq k$, and irreflexive $\forall \lambda_i > k$.

$(ii) \Rightarrow (i)$

- Assume there exist a, b, c, d with

$$\max\{v(a,b), v(c,d)\} < \min\{v(a,d), v(c,b)\} = \lambda,$$

then, S^λ does not satisfy $S^\lambda \left(S^\lambda\right)^d S^\lambda \subset S^\lambda$.

- If there exist a, b, c, d with

$$\max\{v(a,b), v(b,c)\} < \min\{v(a,d), v(d,c)\} = \lambda,$$

then, S^λ does not satisfy $S^\lambda S^\lambda \left(S^\lambda\right)^d \subset S^\lambda$.

Proof of theorem 5.3

$S_i \subset S_{i+1}$, for all i, means that $\forall\, i, j,\ i < j$, we have $aS_i b \Rightarrow aS_j b$. Let i^* be the smallest index such that $aS_i a,\ \forall\, a \in A$. Define

$$v(a,b) = \max\{i^* - i;\ i \text{ such that } aS_i b\}.$$

We have $S_i = S^{\lambda_i}$ with $\lambda_i = i^* - i$; S_i is a strict semiorder iff $i < i^*$ iff $\lambda_i > 0$.

Proof of theorem 5.8

It is easy to see that C is negatively transitive and that condition (i) is equivalent to say that C is asymmetric, so that C is a weak order and $(i) \Longleftrightarrow (ii)$.

We prove now that $(i) \Rightarrow (iii)$. If the line and column entries of $M^{S_{\lambda_i}}$ are ordered according to a complete order $\mathcal{O}$ containing C, we obtain

$$\left(M^{S_{\lambda_i}}\right)_{ac} = 1 \text{ and } \left(M^{S_{\lambda_i}}\right)_{bc} = 0\ \Rightarrow\ \begin{aligned} v(a,c) &> v(b,c) \\ \Rightarrow\quad aCb& \\ \Rightarrow\quad a\mathcal{O}b&, \end{aligned}$$

and

$$\left(M^{S_{\lambda_i}}\right)_{ca} = 0 \text{ and } \left(M^{S_{\lambda_i}}\right)_{cb} = 1\ \Rightarrow\ \begin{aligned} v(c,a) &> v(c,b) \\ \Rightarrow\quad aCb& \\ \Rightarrow\quad a\mathcal{O}b&, \end{aligned}$$

so that $M^{S_{\lambda_i}}$ is a step-type matrix and S_{λ_i} is a semiorder for $\lambda_i \leq k$; it is a strict semiorder for $\lambda_i > k$ and the underlying complete order is the same for all λ_i.

Reciprocally, $(iii) \Rightarrow (i)$ because if $v(c,a) > v(c,b) = \alpha$ and $\beta = v(d,a) > v(d,b)$, then M^{S_α} and M^{S_β} cannot be step-type with the same underlying complete order because

$$\left(M^{S_\alpha}\right)_{ca} < \left(M^{S_\alpha}\right)_{cb} \text{ and } \left(M^{S_\beta}\right)_{da} > \left(M^{S_\beta}\right)_{db}.$$

With the complete order O, the fact that $(M^{S_{\lambda_i}})$ is step-type for every λ_i is possible if and only if M^v is step-type, so that $(iii) \Leftrightarrow (iv)$.

Proof of theorem 5.10

$(i) \Rightarrow (ii)$

As relation C (defined in theorem 5.8) is a weak order, there exists a function g such that, $\forall\, a, b \in A$,

$$aCb \Leftrightarrow g(a) > g(b).$$

Define, $\forall\, a \in A$, $\forall\, \lambda_i \leq k$,

$$g(a) + q_i(a) = \min\{g(c), cS^{\lambda_i}a\},$$

and, $\forall\, a \in A$, $\forall\, \lambda_i > k$,

$$g(a) + q_i(a) = \max\{g(c), c\neg S^{\lambda_i}a\}.$$

The reader will verify, as an exercise, that conditions (ii) are satisfied.

$(ii) \Rightarrow (i)$

Let q be a nonnegative constant and ϵ an arbitrarily small positive number. Indexing the elements of A in such a way that, $\forall\, l$,

$$a_{l-1} \neg\, C\, a_l$$

(this is possible because C is a weak order), we define, $\forall\, a_l \in A$, $\forall\, \lambda_i \leq k$

$$g_i(a_l) = \max\{\max\{g_i(d) - q, a_l S^{\lambda_i} d\}, g_i(a_{l-1}) + \epsilon\},$$

and, $\forall\, a_l \in A$, $\forall\, \lambda > k$,

$$g_i(a_l) = \max\{\max\{g_i(d) + q, a_l S^{\lambda_i} d\}, g_i(a_{l-1}) + \epsilon\},$$

which satisfy condition (iii).
The inverse implications are left to the reader.

Proof of theorem 5.11

If $g_i^*(a) > g_i^*(b)$ for the minimal ϵ-representation g_i^* of S^{λ_i} with threshold q^*, this means that there is $c \in A$ such that $(cS^{\lambda_i}b$ and $c\neg S^{\lambda_i}a)$ or $(aS^{\lambda_i}c$ and $b\neg S^{\lambda_i}c)$. Hence there is c such that $v(a,c) > v(b,c)$ or $v(b,c) > v(c,a)$. Due to property (i) in theorem 5.8, we may not have $g_j^*(a) < g_j^*(b)$ for any $j \neq i$.

Proof of theorem 5.15

$(i) \implies (ii)$. Let x be any fixed element in A. Define $\forall\, a \in A$, $g(a) = v(a,x)$. We have $v(a,b) = v(a,x) - v(b,x) = g(a) - g(b)$. If we denote by λ_i the values of v, ranked in decreasing order, we have

$$
\begin{array}{ll}
aS^{\lambda_i}b & \text{iff} \quad v(a,b) \geq \lambda_i \\
& \text{iff} \quad g(a) - g(b) \geq \lambda_i \\
& \text{iff} \quad g(a) \geq g(b) + \lambda_i.
\end{array}
$$

For $\lambda_i > 0$, S^{λ_i} is a strict semiorder and we can find $\epsilon > 0$ and set $q_i = \lambda_i - \epsilon > 0$ so that

$$aS^{\lambda_i}b \text{ iff } g(a) > g(b) + q_i.$$

For $\lambda_i \leq 0$, S^{λ_i} is a semiorder; we may set $q_i = -\lambda_i$ and have

$$aS^{\lambda_i}b \text{ iff } g(a) \geq g(b) - q_i.$$

$(ii) \implies (i)$ is easier to prove since $[v(a,b) = g(a) - g(b), \ \forall \, a, b \in A]$ implies, $\forall$ $x \in A$,

$$
\begin{aligned}
v(a,b) = g(a) - g(b) &= g(a) - g(x) - (g(b) - g(x)) \\
&= v(a,x) - v(b,x).
\end{aligned}
$$

Proof of theorem 5.17

Suppose first that $(g, q_1, \ldots, q_p)$ is a single-function constant thresholds representation of v. This implies that $(g, q_1, \ldots, q_p, \epsilon)$ is also an ϵ-representation for some $\epsilon > 0$. Let C be any p-cyclone of $\mathcal{P}$. Applying inequality 5.1 to each component circuit of C and summing up all these inequalities yields

$$
\sum_{i=1}^{p} \left(|P_i^d \cap C| - |P_i \cap C| \right) q_i - \epsilon \sum_{i=1}^{p} |P_i \cap C| \geq 0.
$$

If $\displaystyle\sum_{i=1}^{p} |P_i \cap C| = 0$ and C is balanced, then $\displaystyle\sum_{i=1}^{p} |P_i^d \cap C| = 0$ and C is empty, a contradiction. If $\displaystyle\sum_{i=1}^{p} |P_i \cap C| \neq 0$ and C is balanced, the above inequality reduces to $-\epsilon \displaystyle\sum_{i=1}^{p} |P_i \cap C| \geq 0$ which is impossible.

Conversely for proving the existence of an ϵ-representation of v with constant thresholds under the assumption that no p-cyclone is balanced, we may suppose that at least one threshold, say q_p, is not null and through eventual rescaling we may set $q_p = 1$. According to theorem 5.16 sufficient condition for the existence of a constant threshold ϵ-representation is that there is a common solution $q_1, \ldots, q_{p-1}, q_p = 1$ to the system of inequations associated to the set of all nonempty circuits C of $\mathcal{P}$, namely, all inequations of the form

$$
\sum_{i=1}^{p-1} \left(|P_i^d \cap C| - |P_i \cap C| \right) q_i > |P_p \cap C| - |P_p^d \cap C|,
$$

where C is a non-empty circuit of $\mathcal{P}$. Each such inequality determines a half space in $\mathcal{R}^{p-1}$, the space of $(p-1)$ -tuples $(q_1, \ldots, q_{p-1})$. By Helly's theorem (see e.g. Rockafellar 1972, theorem 21.6, p.196), the intersection of all half spaces determined by the above inequalities is non-empty iff any subsystem of p inequations has a solution. So it is sufficient to show that one can find a solution to the system of inequalities associated to any p-cyclone i.e. any non empty p-tuples of circuits from $\mathcal{P}$.

Consider any p-cyclone C composed of circuits $C_1, \ldots, C_p$; let $b_{ij} = |C_i \cap P_j^d| - |C_i \cap P_j|$. The associated system of inequalities reads

$$
(5.3) \quad
\begin{cases}
b_{11} q_1 + b_{12} q_2 + \cdots + b_{1p-1} q_{p-1} > -b_{1p} \\
\ \vdots \qquad \quad \vdots \qquad \ \vdots \quad \vdots \quad \vdots \qquad \quad \vdots \qquad \ \ \vdots \\
b_{p1} q_1 + b_{p2} q_2 + \cdots + b_{pp-1} q_{p-1} > -b_{pp}.
\end{cases}
$$

By theorem 22.2 in Rockafellar 1972, such a system has a solution iff the following implication holds: given real numbers $\lambda_1, \ldots, \lambda_p$, with $\lambda_i \geq 0$ for all i and $\lambda_i > 0$ for at least some i,

$$(5.4) \qquad \left(\sum_{i=1}^{p} \lambda_i b_{ij} = 0, \text{ for } j = 1, 2, \ldots, p-1 \right) \Rightarrow \left(\sum_{i=1}^{p} \lambda_i b_{ip} > 0 \right).$$

Let us restrict the λ_i's to be nonnegative integers. If the implication holds for integers, it is easy to show that it does also for rational and real numbers. For integer λ_i's satisfying the premise of implication 5.4, let us consider the p-cyclone C' composed of the circuits C_i', $i = 1, \ldots, p$ where C_i' consists in passing λ_i times through C_i. In the resulting p-cyclone, there are $\sum_{i=1}^{p} \lambda_i |C_i \cap P_j^d|$ arcs from P_j^d and

$\sum_{i=1}^{p} \lambda_i |C_i \cap P_j|$ arcs from P_j. These number of arcs are equal for $j = 1, 2, \ldots, p-1$

when the premise of the implication 5.4 is satisfied. If C' is not balanced, $\sum_{i=1}^{p} \lambda_i b_{ip}$

cannot be null and we have to prove that it must be positive. Suppose it is strictly negative. This means that there is at least one arc (a, b) of P_p in some C_i; we

build a new p-cyclone C'' by adding $\sum_{i=1}^{p} \lambda_i b_{ip}$ times the loop (a, a) which belongs

to P_p^d; the new p-cyclone C'' is balanced, a contradiction.

6

AGGREGATION OF SEMIORDERS

In this chapter, we have tried to gather some results about the aggregation of a family of semiorders into some "global synthetic structure". Since the subject seems not to have been intensively studied as such, we have tried to formulate a general framework which encompasses a number of well-known methods of preference aggregation. By "aggregation" we generally mean a "mechanism" which associates some structure to any family of structures of a given type (semiorders, for instance) in a "sensible" manner. Of course, what is "sensible" takes different forms according to the context; here we focalize on preference modelling and decision situations. Note that the results we state below can very often be extended to interval orders; this will be eventually mentioned.

We start by examining the general problem of aggregating complete preorders into a semiorder and we state an Arrow-like impossibility result; of course the impossibility of aggregating semiorders into a semiorder or a complete preorder follows as a by-product.

So, having shown that the problem might be uneasy, we deal with the simplest scheme for aggregating ordinal structures, namely the lexicographic procedure. What is essentially shown is the difficulty of defining a satisfactory lexicographic method for semiorders, although it is immediate for complete preorders. Difficulties also arise with the generalization of another classic of orders aggregation, Borda's sum of ranks method; there, the problems come from the fact that there is no widely accepted definition of the rank for semiorders; we however make a proposal which is a product of the deep investigations into the semiorder structure exposed in Chapter 4.

Anyway, the above results show that the schemes that work well for complete orders or preorders do not lean themselves easily to generalization. That is why we devote the rest of the chapter to developing a general framework for the aggregation of semiorders or interval orders (encompassing of course the case of complete orders and preorders). Two main types of methods can fit into this framework. The first one can be considered as inspired by the utility paradigm; it consists in the aggregation of numerical representations into a "global evaluation" function, through weighted sums or more general operators. The second type of methods derives from the idea of pairwise comparisons à la Condorcet; it encompasses some of the so-called outranking methods like versions of ELECTRE. We present very

general conditions allowing to describe a global preference structure as the result
of some operation performed on a family of semiorders. In the final section, we
extend this formalism to deal with valued relations, encompassing then outranking
methods such as PROMETHEE.

6.1   Arrow's theorem for semiorders

Let A be a finite set, n an integer ≥ 3, X a set of n-tuples of relations on A and
Y a set of relations on A.

Definition 6.1
*(i) f is an (X, Y)-aggregation procedure if it is a mapping from X to Y, which
associates, to every element $(S_1, S_2, ..., S_n)$ of X, an element $S = f(S_1, S_2, ..., S_n)$
of Y;*
(ii) f satisfies the unanimity *rule if, $\forall (S_1, S_2, ..., S_n) \in X, \forall a, b \in A$,*

$$a \, P_i \, b, \ \forall i \implies a \, P \, b,$$

where P_i and P are the asymmetric parts respectively of S_i and $S = f(S_1, S_2, ..., S_n)$;
(iii) f satisfies the non-dictatorsphip *rule if $\nexists i \in \{1, 2, ..., n\}$ such that,
$\forall (S_1, S_2, ..., S_n) \in X, \forall a, b \in A$,*

$$a \, P_i \, b \implies a \, P \, b;$$

(iv) f satisfies the independence *rule if,
$\forall (S_1, S_2, ..., S_n), (S_1', S_2', ..., S_n') \in X, \forall a, b \in A$,
$(S_1, S_2, ..., S_n)/\{a, b\} = (S_1', S_2', ..., S_n')/\{a, b\} \implies S/\{a, b\} = S'/\{a, b\}$,
where $S = f(S_1, S_2, ..., S_n)$, $S' = f(S_1', S_2', ..., S_n')$,
$S/\{a, b\}$ is the restriction of S to $\{a, b\}$ and
$(S_1, S_2, ..., S_n)/\{a, b\} = (S_1/\{a, b\}, S_2/\{a, b\}, ..., S_n/\{a, b\})$.*

Theorem 6.1 *If $|A| \geq 4$, if X is the set of all n-tuples of complete preorders on
A and if Y is the set of all semiorders on A, then there is no (X, Y)-aggregation
procedure satisfying simultaneously the unanimity, non-dictatorship and independence rules.*

The proof is given in section 6.9.

Comments

• The cases where X is the set of all n-tuples of semiorders on A and/or Y is the
set of all complete preorders on A are immediate corollaries of this theorem; note
that in this case, the theorem is valid as soon as $|A| \geq 3$

• If all the considered relations are complete preorders, theorem 6.1 is exactly
Arrow's theorem for $|A| \geq 4$ (Arrow 1963). It is well-known that Arrow's theorem

is also true for $|A| = 3$ but this is no longer verified when the elements of Y are semiorders. For example, the procedure which consists in defining

$$\left\{ \begin{array}{ll} a\,P\,b & \text{iff } a\,P_i\,b,\ \forall i, \\ a\,I\,b & \text{otherwise,} \end{array} \right.$$

always provides a semiorder when $|A| = 3$ and satisfies simultaneously the unanimity, the non-dictatorship and the independence rules.

• As the reader will see in the proof of the theorem, only the "$PIP \subset P$" property of the semiorders is used, so that the theorem is still true when Y is the set of interval orders on A.

6.2 Lexicographic aggregation of semiorders

This section is essentially based on Pirlot and Vincke 1992.

6.2.1 The usual rule

Let $\{ (P_j, I_j), j = 1, 2 ..., n \}$ be a family of complete preorders numbered in decreasing order of importance. The usual lexicographic rule consists of defining the preference relation $P^{(0)}$ and the indifference relation $I^{(0)}$ as follows:

$$\left\{ \begin{array}{ll} a\,P^{(0)}\,b & \text{iff } \exists k: a\,P_k\,b \text{ and } a\,I_j\,b,\ \forall j < k, \\ a\,I^{(0)}\,b & \text{iff } a\,I_j\,b,\ \forall j. \end{array} \right.$$

It is well known (see Fishburn 1974) that in this case the obtained structure $(P^{(0)}, I^{(0)})$ is a complete preorder.

The introduction of indifference thresholds completely modifies this result; indeed, if $\{ (P_j, I_j), j = 1, 2 ..., n \}$ is a family of semiorders numbered in decreasing order of importance, the relation $P^{(0)}$ obtained by the usual rule may contain cycles as illustrated by example 1.

Example 1 $A = \{ a, b, c \}; n = 2$; see tables 6.1 and 6.2.

1	a	b	c
a		P	
b			
c			

Table 6.1

2	c	b	a
c		P	P
b			P
a			

Table 6.2

The usual lexicographic rule leads to $bP^{(0)}a$, $aP^{(0)}c$, $cP^{(0)}b$.

It can be argued that the usual rule does not exploit the whole information provided by the semiorders. In the previous example, for instance, the fact that a "has an advantage" over b in the first semiorder (a is preferred to c while b is not) is completely neglected by the usual rule. Thus it may be interesting to investigate rules which explicitly take into account "advantages" of some actions over some others.

Remark

Using a numerical representation $g_i, i = 1, 2, ..., n$ of the given structures, the usual lexicographic rule can be rewritten as

$$a\ P^{(0)}\ b \text{ iff } (g_1(a), g_2(a), ...g_n(a)) >_L (g_1(b), g_2(b), ...g_n(b))$$

for complete preorders, and as

$$a\ P^{(0)}\ b \text{ iff } (g_1(a), g_2(a), ...g_n(a)) >_L (g_1(b), g_2(b), ...g_n(b)) + (q_1, q_2, ...q_n)$$

for semiorders, where $>_L$ denotes lexicographic ordering on R^n and $(q_1, q_2, ...q_n)$ is a vector of positive constants.

6.2.2   First variant

We denote by (P'_j, I'_j) the complete preorder associated with semiorder (P_j, I_j) (see remark in section 3.16). We complement the usual rule by requiring that the first semiorder giving an advantage and the first semiorder giving a strict preference, when different, do not contradict each other;

$$\begin{aligned} aP^{(1)}b \quad &\text{iff } \exists k : aP_k b \text{ and } aI_j b, \forall j < k \\ &\text{and } \exists l \le k : a\ P'_l\ b \text{ and } a\ I'_j\ b, \forall j < l. \end{aligned}$$

Theorem 6.2 $P^{(1)}$ *has no cycle.*

The proof is given in section 6.9.

Theorem 6.3 $P^{(1)}$ *is transitive for $n = 2$ but not necessarily for $n > 2$.*

The proof of the first assertion is given in section 6.9. Examples 2 and 3 illustrate the second assertion.

Example 2 $A = \{ a, b, c, d, e \}; n = 4$; see tables 6.3 to 6.6.

In that example, we have $a\ P^{(1)}\ b$, $b\ P^{(1)}\ c$, $a\ \neg P^{(1)}\ c$, $c\ \neg P^{(1)}\ a$.
    The fact that $P^{(1)}$ has no cycle is interesting in itself because it is easier to choose the best actions or to rank the actions on the basis of an acyclic preference relation. However, if we want a transitive relation, it is necessary to strengthen the first variant.

1	a	b	c	d	e
a			P	P	
b				P	
c					
d					
e					

Table 6.3

2	c	b	a	d	e
c				P	P
b					P
a					
d					
e					

Table 6.4

3	b	a	c	d	e
b			P	P	P
a				P	P
c				P	P
d					
e					

Table 6.5

4	c	a	b	d	e
c		P	P	P	P
a			P	P	P
b					
d					
e					

Table 6.6

6.2.3 Second variant

One manner of strengthening the first variant is to require that the advantages in favour of a "cover" the advantages in favour of b, i.e.

$a\ P^{(2)}\ b$ iff $\exists k : aP_k b$ and $aI_j b, \forall j < k$

and there is a function $\phi : J \to L$ with

$J = \{\, j \in \{1, 2, ..., n\}$ such that $j < k$ and $b\ P'_j\ a\}$,

$L = \{\, l \in \{1, 2, ..., n\}$ such that $a\ P'_l\ b\}$,

such that $\begin{cases} \forall j \in J,\ \phi(j) < j, \\ \forall j, j' \in J,\ j \neq j' \text{ implies } \phi(j) \neq \phi(j'). \end{cases}$

Thus, for each semiorder giving an advantage to b over a, there exists a more important semiorder giving an advantage to a over b.

Theorem 6.4 $P^{(2)}$ *has no cycle.*

This theorem is an immediate consequence of theorem 6.3 and of the obvious fact that $P^{(2)} \subset P^{(1)}$.

Theorem 6.5 $P^{(2)}$ *is transitive for* $n = 2$ *but not necessarily for* $n > 2$.

The proof for $n = 2$ is the same as in theorem 6.3 (in this case $P^{(1)} = P^{(2)}$). The following example illustrates the second part of the theorem.

Example 3 $A = \{\, a, b, c, d, e\,\}; n = 3$; see tables 6.7 to 6.9.
This leads to $a\ P^{(2)}\ b$, $b\ P^{(2)}\ c$, $a\ \neg P^{(2)}\ c$, $c\ \neg P^{(2)}\ a$.

1	a	b	c	d	e
a				P	P
b					P
c					
d					
e					

Table 6.7

2	a	c	b	d	e
a			P	P	P
c					
b					
d					
e					

Table 6.8

3	b	c	a	d	e
b		P	P	P	P
c			P	P	P
a					
d					
e					

Table 6.9

6.2.4 Third variant

The previous strengthening does not lead to transitivity; the previous example illustrates the fact that the presence of any semiorder giving an advantage in contradiction with the "desired" preference is sufficient to destroy transitivity, so that it seems necessary to adopt the following rule if transitivity is wanted. We ask now that no semiorder which is more important than the first semiorder yielding a strict preference gives an advantage in contradiction with it:

$$a\, P^{(3)}\, b \text{ iff } \exists k: \ a\, P_k\, b \text{ and } a\, (P_j' \cup I_j')\, b, \ \forall j < k.$$

Theorem 6.6 $P^{(3)}$ *is transitive but not negatively transitive.*

See section 6.9 for a proof of the transitivity property. The second part of the theorem is illustrated by the following example.

Example 4 $A = \{a, b, c\}; n = 2$; see tables 6.10 and 6.11.

1	a	b	c
a		P	
b			
c			

Table 6.10

2	c	b	a
c		P	P
b			P
a			

Table 6.11

We have $a \, \neg P^{(3)} \, b$, $b \, \neg P^{(3)} \, a$, $b \, \neg P^{(3)} \, c$, $c \, \neg P^{(3)} \, b$, $a \, P^{(3)} \, c$.

6.2.5 A note on strict preference

It is straightforward to obtain an acyclic preference relation, but for transitivity, we need a much stronger rule. Moreover, none of the previous rules would give a preference of a over b in the following example, where such a preference would be rather intuitive.

Example 5 $A = \{a, b, a_i, b_j; i = 1, 2, ..., 8; j = 1, 2, ..., 10\}; n = 2$; see tables 6.12 and 6.13.

1	b	a	a_1	....	a_8	b_1	....	b_{10}
b		I	I	....	I	P	....	P
a			I	....	I	I	$P...$	P
a_1				$I...$	I	I	$IP..$	P
.								
.								
.								
a_8						$I...I$	P	
b_1						$I..$	I	
.								
.								
.								I
b_{10}								

Table 6.12

2	a	a_1	...	a_8	b_1	...	b_{10}	b
a		P	....					P
a_1			$P.......$					P
.								
.								
a_8					$P...$			P
b_1					$I...$			I
.								
b_{10}								I
.								.
.								.
b								.

Table 6.13

In this example the first semiorder gives an advantage in favour of b over a, which could be qualified as "weak" because, even cumulated with advantages in favour of a over a_1 and in favour of a_i over $a_{i+1}, \forall i$, it still yields an indifference between b and a_8. On the other hand, the second semiorder gives a strict preference of a over b which could be qualified as "strong" because of the chain of strict preferences between a and b.

This example points out a more fundamental question: the structure of semiorder (the presence of an indifference threshold) implicitly conveys information on the "degree" of indifference and of preference between the actions (which is formalized by the concept of minimal representation, in chapter 4). The question is whether it is possible to exploit this information in our aggregation problem, avoiding, however, a complete "cardinalization" of the situation.

6.2.6   A note on indifference

It is easy to get a "nice" indifference relation by defining

$$a \; I \; b \text{ iff } a \; I'_j \; b, \; \forall j.$$

It is nice in the sense that it is an equivalence relation (reflexive, symmetric and transitive); combined with $P^{(3)}$ it gives a individual preorder, but of course this indifference relation will in general be very sparse.

The use of $I^{(0)}$ given by the usual rule may also be considered as natural but does not lead to very nice properties, even when combined with $P^{(3)}$: the following example shows that $(P^{(3)}, I^{(0)})$ is not even a individual interval order (see Roubens and Vincke 1985, page 54).

Example 6 $A = \{a, b, c, d\}; n = 2$; see tables 6.14 and 6.15.

1	a	c	d	b
a				P
c				
d				
b				

2	b	c	d	a
b			P	P
c			P	P
d				
a				

        Table 6.14   Table 6.15

We have $a \; P^{(3)} \; b$, $b \; I^{(0)} \; c$, $c \; P^{(3)} \; d$ but $a \; I^{(0)} \; d$.

The two previous definitions lead to individual preference structures (some pairs of actions may remain incomparable). If we wish to avoid incomparability, we may define indifference as an absence of strict preference as follows (we only consider the case of $P^{(3)}$):

$$a \; I^{(3)} \; b \text{ iff } a \; \neg P^{(3)} \; b \text{ and } b \; \neg P^{(3)} \; a.$$

This relation is certainly not transitive (we showed that $P^{(3)}$ is not negatively transitive); moreover, $(P^{(3)}, I^{(3)})$ is not even an interval order (take again the latter example where $I^{(0)}$ may be replaced by $I^{(3)}$).

6.2.7 Fourth variant

Variants 4 and 5 below are concerned with the case where a semiorder is requested as aggregated structure. The rules obtained are fairly sophisticated and we leave it to the reader to appreciate their lexicographic character.

We first consider the case where $n = 2$ and we adopt the following rule; a is preferred to b if

$$\left\{ \begin{array}{l} a \text{ is preferred to } b \text{ for the first semiorder} \\ \text{or} \\ a \text{ is equivalent to } b \text{ for the first semiorder and preferred for the second one.} \end{array} \right.$$

A third case has to be added in order to get a semiorder; a is also preferred to b if

$$\left\{ \begin{array}{l} a \text{ has an advantage over } b \text{ for the first semiorder and there exist } c, d \\ \text{``between'' } a \text{ and } b \text{ such that } c \text{ is globally preferred to } d \text{ for the} \\ \text{aggregated relation.} \end{array} \right.$$

More formally,

$$\begin{array}{rl} a\, P^{(4)}\, b & \text{iff} \quad a\, P_1\, b \\ & \text{or} \quad a\, I_1'\, b \text{ and } a\, P_2\, b \\ & \text{or} \quad a\, I_1\, b,\, a\, P_1'\, b \text{ and } \exists\, c, d \text{ such that} \\ & \qquad a\, (P_1' \cup I_1')\, c\, I_1'\, d\, (P_1' \cup I_1')\, b \text{ and } c\, P_2\, d. \end{array}$$

Theorem 6.7 $(P^{(4)}, I^{(4)})$ *is a semiorder, where*

$$a\, I^{(4)}\, b \text{ iff } a\, \neg P^{(4)}\, b \text{ and } b\, \neg P^{(4)}\, a.$$

The proof can be found in section 6.9.

Remark

From the proof of this theorem we know that $P^{(4)'} \subseteq P'$ and $I^{(4)'} \supseteq I'$ because the matrix associated with $P^{(4)}$ is an upper diagonal step-type matrix when the elements of A are ranked according to (P', I'). In other words the complete preorder associated with $(P^{(4)}, I^{(4)})$ is contained in the complete preorder obtained by the application of the usual lexicographic rule to the complete preorders associated with the two given semiorders. In general the two complete preorders are not equal, as can be seen from the example below.

The aggregation of n semiorders (with $n > 2$) is performed by a recursive application of the previous rule as follows:

Step 1 Apply the rule to $\{(P_1, I_1), (P_2, I_2)\}$ to obtain the semiorder $(P_{\underline{2}}, I_{\underline{2}})$;

Step k Apply the rule to $\{ (P_{\underline{k}}, I_{\underline{k}}), (P_{k+1}, I_{k+1}) \}$ to obtain the semiorder $(P_{\underline{k+1}}, I_{\underline{k+1}})$; $k = 2, 3, ..., n - 1$.

1	a	b	c	d	e	f	g
a					P	P	P
b					P	P	P
c							
d							
e							
f							
g							

Table 6.16

2	g	c	d	a	b	e	f
g				P	P	P	P
c				P	P	P	P
d				P	P	P	P
a							P
b							
e							
f							

Table 6.17

3	b	a	d	f	g	c	e
b		P	P	P	P	P	P
a				P	P	P	P
d						P	P
f						P	P
g						P	P
c							
e							

Table 6.18

By recursive application of theorem 6.7, the final structure is a semiorder.

Example 7 $A = \{\, a, b, c, d, e, f, g \,\}$; $n = 3$; see tables 6.16, 6.17 and 6.18.

Steps 1 and 2 of the iterative application of the rule yield the semiorders respectively defined by tables 6.19 and 6.20.

$P_2^{(4)}$	a	b	c	d	g	e	f
a					P	P	P
b					P	P	P
c						P	P
d						P	P
g						P	P
e							
f							

Table 6.19

$P_3^{(4)}$	b	a	d	c	g	f	e
b		P	P	P	P	P	P
a				P	P	P	P
d				P	P	P	P
c						P	P
g						P	P
f							P
e							

Table 6.20

Note that $(P_2^{(4)'}, I_2^{(4)'})$ is not the lexicographic aggregation of (P'_1, I'_1) and (P'_2, I'_2); we have $a I_2^{(4)'} b$ while $a I'_1 b$ and $a P'_2 b$. A fortiori, $(P_3^{(4)'}, I_3^{(4)'})$ is not the lexicographic aggregation of the complete preorders associated with the three given semiorders.

6.2.8 Fifth variant

By slightly modifying the previous rule, one can obtain an aggregated relation which is a semiorder whose associated complete preorder is the lexicographic aggregation of the complete preorders (P'_j, I'_j), $j = 1, 2, ..., n$. For two semiorders,

$$a \; P^{(5)} \; b \quad \text{iff} \quad a \; P_1 \; b$$
$$\text{or} \quad a \; I'_1 \; b \text{ and } a \; P'_2 \; b$$
$$\text{or} \quad a \; I_1 \; b, \; a \; P'_1 \; b \text{ and } \exists c, d \text{ such that}$$
$$a \; (P'_1 \cup I'_1) \; c \; I'_1 \; d \; (P'_1 \cup I'_1) \; b \text{ and } c \; P'_2 \; d.$$

The only difference from the fourth variant is the replacement of P_2 by P'_2, which of course yields a richer strict preference at least for $n = 2$. Defining $I^{(5)}$ in the usual way, we have the following theorem.

Theorem 6.8 $(P^{(5)}, I^{(5)})$ *is a semiorder and* $(P^{(5)'}, I^{(5)'})$ *is the complete preorder* (P', I') *obtained by applying the lexicographic rule to the complete preorders* (P_1', I_1') *and* (P_2', I_2').

The proof is given in section 6.9.

The aggregation of n semiorders (with $n > 2$) will be made by recursive application of the rule. Steps 1 and 2 of the iterative application of the rule to the last example yield the semiorders in tables 6.21 and 6.22

$P_2^{(5)}$	a	b	c	d	g	e	f
a		P	P	P	P	P	P
b				P	P	P	P
c					P	P	P
d					P	P	P
g					P	P	P
e							P
f							

$P_3^{(5)}$	a	b	d	c	g	e	f
a		P	P	P	P	P	P
b				P	P	P	P
d				P	P	P	P
c						P	P
g						P	P
e							P
f							

Table 6.21   Table 6.22

Final remark

From a practical viewpoint, we would in general recommend the fourth variant rather than fifth one. In effect, the latter, though appealing owing to its additional property, is likely to "add" more preferential information (as illustrated by the example below). In a further study it could be of interest to look for a rule which could be qualified as lexicographic and lead to a complete semiorder with an asymmetric part as small as possible (for instance, in the sense of inclusion).

Example 8 $A = \{a, b, c, d, e, f\}; n = 2$; see tables 6.23 and 6.24.

Application of variants 4 and 5 yield the semiorders in tables 6.25 and 6.26 respectively.

1	a	b	c	d	e	f
a						P
b						
c						
d						
e						
f						

2	b	c	d	f	e	a
b					P	P
c						P
d						
f						
e						
a						

Table 6.23   Table 6.24

$P^{(4)}$	a	b	c	d	e	f
a					P	P
b				P	P	
c						
d						
e						
f						

$P^{(5)}$	a	b	c	d	e	f
a			P	P	P	P
b			P	P	P	P
c				P	P	P
d					P	P
e						
f						

Table 6.25   Table 6.26

6.3   Aggregation of semiorders by Borda's method

6.3.1   Introduction

We first recall, on the basis of an example, how Borda's method works for the aggregation of strict complete orders.

Consider 4 elements a, b, c, d and the 3 following strict complete orders:

$$a \ P_1 \ b \ P_1 \ c \ P_1 \ d,$$
$$c \ P_2 \ a \ P_2 \ d \ P_2 \ b,$$
$$b \ P_3 \ c \ P_3 \ d \ P_3 \ a.$$

Borda's method consists in ranking the elements in decreasing order of the sum of their *ranks*, yielding

$$\text{for } a, \ 3 \ + \ 2 \ + \ 0 \ = \ 5,$$
$$\text{for } b, \ 2 \ + \ 0 \ + \ 3 \ = \ 5,$$
$$\text{for } c, \ 1 \ + \ 3 \ + \ 2 \ = \ 6,$$
$$\text{for } d, \ 0 \ + \ 1 \ + \ 1 \ = \ 2,$$

hence

$$cP\{a,b\}Pd.$$

In the case of strict complete orders, it is equivalent to rank the elements in the decreasing order of the sum of their *scores*, where the *score* of an element x is the difference between the number of elements which are "worse than" x and the number of elements which are "better than" x. In our example, we obtain

$$
\begin{array}{llllllll}
\text{for a,} & (3-0) & + & (2-1) & + & (0-3) & = & 1, \\
\text{for b,} & (2-1) & + & (0-3) & + & (3-0) & = & 1, \\
\text{for c,} & (1-2) & + & (3-0) & + & (2-1) & = & 3, \\
\text{for d,} & (0-3) & + & (1-2) & + & (1-2) & = & -5.
\end{array}
$$

The equivalence between both methods is due to the fact that, in a strict complete order, the score $S(x)$ and the rank $R(x)$ of an element x are linked by

$$S(x) = 2R(x) - (n-1),$$

where n is the number of elements.

Many different axiomatisations of Borda's method have been established, for instance by Young 1974, Hansson and Sahlquist 1976, Nitzan and Rubinstein 1981, Marchant 1996, to cite only a few authors.

There is no classical concept of rank of an element in a semiorder and the purpose of this section is to indicate various possible ways: the goal is of course to take into account, as much as possible, the particular structure of semiorders and the information they contain. None of the concepts of rank we present is fully satisfactory, but there are arguments pleading for the one we introduced in Section 4.7 (these are exposed in subsection 6.3.4). To illustrate the possible ways, we will consider the aggregation of the four semiorders represented in figure 6.1.

6.3.2 Use of the concept of rank of an element in a graph without circuit

The rank of a node in an oriented graph without circuit is the length of the longest path issued from that node (note that the usual definition is rather stated in terms of longest paths ending up in the node; the former definition obviously corresponds to the usual notion when applied to the graph obtained by reversing the orientations of all arcs). As the asymmetric part of a semiorder can be represented by a graph without circuit, we can use this concept and rank the elements in decreasing order of the sum of their ranks. This gives, in the example of figure 6.1

$$
\begin{array}{llllllllll}
\text{for a,} & 1 & + & 0 & + & 0 & + & 0 & = & 1, \\
\text{for b,} & 0 & + & 0 & + & 1 & + & 1 & = & 2, \\
\text{for c,} & 1 & + & 0 & + & 0 & + & 0 & = & 1, \\
\text{for d,} & 0 & + & 1 & + & 0 & + & 0 & = & 1.
\end{array}
$$

This method does not discriminate very much the elements; in particular, it does not take into account the number of elements which are "worse" than a given element; it considers as equivalent, for example, the position of c in the first semiorder and the position of d in the second one. The exploitation of the information contained in the semiorders is not really satisfactory.

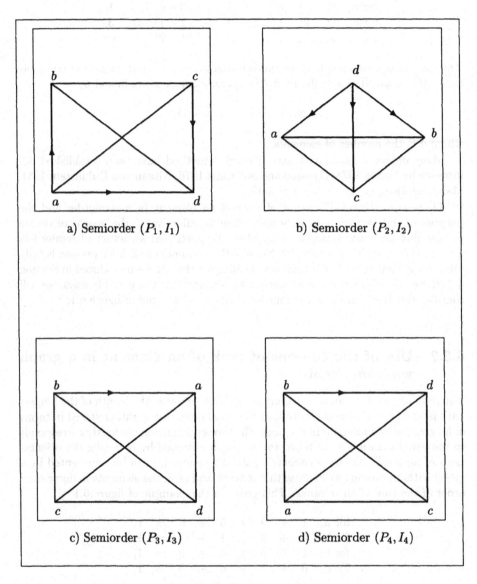

a) Semiorder (P_1, I_1)

b) Semiorder (P_2, I_2)

c) Semiorder (P_3, I_3)

d) Semiorder (P_4, I_4)

Figure 6.1: Four semiorders

6.3.3   Using scores

Applying the method of scores (as presented in the introduction) to the asymmetric
part of the semiorders in figure 6.1 gives the following result:

$$
\begin{array}{lcccccccl}
\text{for } a, & (2-0) & + & (0-1) & + & (0-1) & + & (0-0) & = & 0, \\
\text{for } b, & (0-1) & + & (0-1) & + & (1-0) & + & (1-0) & = & 0, \\
\text{for } c, & (1-0) & + & (0-1) & + & (0-0) & + & (0-0) & = & 0, \\
\text{for } d, & (0-2) & + & (3-0) & + & (0-0) & + & (0-1) & = & 0.
\end{array}
$$

As the previous one, this method is not discriminant enough; all strict preferences
are treated in the same manner and the information contained in the indifference
relations is not taken into account. To show the latter point, it is interesting to
determine the complete preorders associated with the semiorders and to represent
each semiorder together with its underlying complete preorder (the elements are
placed from top to bottom on the basis of the complete preorder, with all equivalent
elements on the same level), as in figure 6.2.

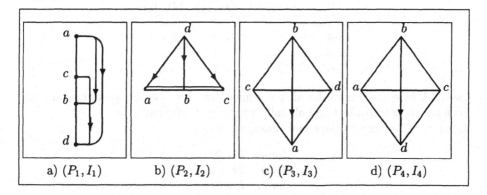

Figure 6.2: Another representation of the semiorders of figure 6.1

  We see in this representation that, in a certain way, the preference of a over
d in (P_1, I_1) can be considered as "stronger" than the preference of d over a in
(P_2, I_2). So, it can be interesting to define a concept of *rank* on the basis of the
paths of preferences and indifferences.

6.3.4   Generalization of Borda's method

In this section, we use in an intuitive manner, a generalization to the semiorders,
of the concept of rank; it is observed by the end of this section that this concept
coincides with the one already presented in section 4.7.
  In a complete order, the rank of an element can be defined as the number of
elements which are "worse" than it is or, in the associated graph, the length of the
longest path issued from that element. In a semiorder, the graph contains P-arcs
and I-arcs; we associate values to these arcs and consider paths of maximal *value*.

As an example, let us envisage the case where the value of a P-arc is considered as slightly greater than the value of an I-arc, by associating the value $(k + 1)$ to P and $(-k)$ to I, where $k \geq 1$. If p and i are respectively the numbers of P's and I's of a path, the value of this path will be

$$p(k + 1) - ik = vk + p,$$

where $v = p - i$.

We define the rank of an element as the maximal value of a path issued from that element. In the case of (P_1, I_1), we obtain

$$
\begin{aligned}
\text{rank of } a &= k + 2 \quad \text{(path } abcd\text{)}, \\
\text{rank of } b &= 1 \quad\quad\,\, \text{(path } bcd\text{)}, \\
\text{rank of } c &= k + 1 \quad \text{(path } cd\text{)}. \\
\text{rank of } d &= 0.
\end{aligned}
$$

The other semiorders give the following results:

	(P_2, I_2)	(P_3, I_3)	(P_4, I_4)
rank of a	0	0	1
rank of b	0	$k+1$	$k+1$
rank of c	0	1	1
rank of d	$k+1$	1	0

(we took here the same k for all the semiorders; this means that we consider all preferences as equivalent and all indifferences as equivalent).

Computing the sum of these new ranks, we obtain

$$
\begin{aligned}
\text{for } a, &\quad k + 3, \\
\text{for } b, &\quad 2k + 3, \\
\text{for } c, &\quad k + 3, \\
\text{for } d, &\quad k + 2,
\end{aligned}
$$

hence the complete preorder

$$bP\{a, c\}Pd.$$

In this example, the final complete preorder does not depend on the value of k. More generally, the complete preorder will be the same for every sufficiently large value of k, as the rank is of the form $vk + p$ (v will determine the value of the rank and p will make the difference between elements with the same value of v).

Comments

The axiomatization of this generalized Borda's method is still an open problem. There are however a number of arguments pleading for using the notion of rank suggested above. The most interesting one is the following and it relies on rather deep properties from Chapter 4. For each preference relation (P_i, I_i), the rank of an element can be expressed by a formula of the type $v_i k + p_i$, where p_i is the number of P_i-arcs and $v_i + p_i$ is the number of I_i-arcs of a maximal value path issued from the element (v_i is positive since there are more I_i- than P_i-arcs in any circuit). It is not clear a priori that such paths remain maximal for all values of k; however, according to theorem 4.20, at least one path from each element remains maximal for all values of k compatible with an ε-representation (with $\varepsilon = 1$, here) of the semiorder; those paths are the maximal *length* paths in a partial graph of the graph associated to the semiorder. This implies that v_i and p_i can be viewed as constants, relatively to k; hence, our definition of rank is linked with geometric properties of the graph, not on particular choices of values associated to the arcs P_i and I_i.

The connection with the minimal representation of semiorders is straightforward as mentioned in section 4.7. The ranks considered here are the minimal ε-representations with threshold k. If the value of k is chosen to be as small as possible (i.e. $k = \alpha$, see theorem 4.2), the obtained rank is the minimal ε-representation (with $\varepsilon = 1$), i.e. the smallest representation of the semiorder on the integers. However, in the aggregation context, we may not take the minimal representation for all (P_i, I_i) since the minimal values of their respective thresholds are not equal in general.

We conclude this discussion by making two additional remarks. Instead of defining the rank on the basis of maximal paths issued from each element, we could also consider maximal paths finishing in each element. For complete orders, this corresponds to counting the number of elements which are better than a given one; it yields another rank function which is just the total number of elements minus the former rank; both ranks thus yield the same ranking when used in Borda's method; this is not the case in general (see the comments after theorem 4.20), although both concepts are equally well justified.

Finally note that with our generalization of Borda's method, one could also consider the global score as defining a semiorder (e.g. with threshold k) instead of a complete preorder.

6.4 The dominant aggregation paradigm applied to semiorders

The most widespread paradigm for synthesizing multidimensional information about a set of alternatives consists in aggregating them into a one-dimensional overall evaluation on a numerical scale; the overall evaluation is obtained by combining numerical encodings of the one-dimensional information. In the dominant models of decision-making and economic theory the overall preference relation $\succeq$ is a complete preorder and the strict overall preference $\succ$ is a weak order on the set of alternatives. Many papers have been devoted to establishing conditions under which this overall preference can be numerically represented by a function u such that, $\forall\, a, b \in A$,

$$a \succeq b \text{ iff } u(a) \geq u(b)$$

and the overall "utility" (or value) u is a function of individual "utilities" u_i, $i = 1, \ldots, n$, on each point of view, i.e.

$$u(a) = f(u_1(a), \ldots, u_n(a)).$$

In particular, in the additive utility model or additive conjoint representation (first axiomatized in Debreu 1960; see Krantz, Luce, Suppes and Tversky 1978, p.256 et sq., for historical comments),

$$u(a) = \sum_{i=1}^{n} u_i(a).$$

The latter model of course results in complete and transitive overall preferences excluding or at least neglecting the possibility of semiordered individual preferences. A number of experimental studies (May 1954, Tversky 1969, MacCrimmon and Larsson 1979) have shown that preorders or weak orders are not in all cases adequate as models for real-life preferences. In particular the importance of models for intransitive indifference has been stressed (Luce 1956)). In spite of criticisms on both empirical and theoretical grounds, the complete preorder model for preferences remains largely dominant and little effort has been devoted to the study of more general aggregation schemes (see Fishburn 1991b for a survey of models for intransitive preferences).

In the tradition of the literature about preference structures, either a descriptive or a normative viewpoint is generally adopted (these viewpoints are extensively discussed in Bell, Raiffa and Tversky 1988). This postulates that in some sense preferences are a well-defined object of investigation, have a form of existence even if they may sometimes be hidden. The role of a decision-making process is to reveal these existing preferences by computing them from more elementary informations about individual preferences, tradeoffs, etc. We adopt here a more constructive viewpoint, in line with the work of B. Roy (Roy 1985, Roy and Bouyssou 1993; see also Roy 1993, for a presentation of the constructive approach). In the latter perspective, overall preferences do not pre-exist the interactive decision building process; they are constructed in the course of a learning process which does not often lead to preference relations with nice properties; they may lack completeness, acyclicity, transitivity of indifference or, even, of strict preference.

The constructive approach is also in line with the natural attitude which consists in building up an overall evaluation of objects which are described on several viewpoints. The most common illustration of this attitude is the use of weighted sums of individual evaluations as is used e.g. for evaluating students at school. We tend to consider utility theory as a rationalization of this "basic instinct". Anyway, this approach can be adapted to deal with semiordered individual information. Consider the following situation in which the semiorder structure receives its most direct interpretation.

Let A be a finite set of alternatives which are evaluated w.r.t. n criteria. Each alternative $a \in A$ is measured (evaluated) on the ith criterion by a value $g_i(a)$. The measurement imprecision on the ith scale is q_i which means that one is unable to discriminate (and hence to formulate preferences) between alternatives differing by less than q_i on the ith criterion. Suppose that all individual preferences are increasing with g_i's. Hence we have, $\forall\, a, b \in A$,

$$a \succ_i b \quad \text{iff} \quad g_i(a) > g_i(b) + q_i$$

and

$$a \sim_i b \quad \text{iff} \quad |g_i(a) - g_i(b)| \leq q_i,$$

where $\succ_i$ (resp. $\sim_i$) denotes the strict preference (resp. indifference) on the ith criterion. All $\succ_i$'s are strict semiorders. A simple way of building an overall binary preference relation is by means of numerical relations:

$$\begin{cases} a \succ b & \text{iff} \quad f(a) > f(b) + q, \\ a \sim b & \text{iff} \quad |f(a) - f(b)| \leq q, \end{cases}$$

$$\text{with} \quad \begin{cases} f(a) = \displaystyle\sum_{i=1}^{n} w_i g_i(a), \\ q = \displaystyle\sum_{i=1}^{n} w_i q_i, \end{cases} \qquad \text{(model } \mathbf{M_1}\text{)}$$

and w_i, $i = 1, \ldots, n$, non-negative real numbers. The resulting overall preference is then a semiorder. There is a straightforward interpretation for f in case the g_i's *measure* the alternatives with some imprecision q_i on dimension i and there is an objective tradeoff, constant along the scale, for converting a measure along i in a measure along any other dimension. In view of the choice of q, this aggregation procedure could be said "prudent" as, if we have a, b with a slightly preferred to b on criterion 1 for instance, and "nearly" preferred but still indifferent to b on all remaining viewpoints, we could declare a globally indifferent to b. To be more precise, let

$$\begin{cases} g_1(a) = g_1(b) + q_1 - \frac{\varepsilon}{w_1}, \\ g_i(a) = g_i(b) + q_i + \frac{\varepsilon}{(n-1)w_i}, \end{cases} \qquad i = 2, 3, \ldots, n,$$

ε being a sufficiently small positive number; then

$$f(a) = \sum_{i=1}^{n} w_i g_i(a) = \sum_{i=1}^{n} w_i g_i(b) + \sum_{i=1}^{n} w_i q_i = f(b) + q$$

and we conclude that $a \sim b$. In other words, the overall indifference threshold in this model is rather large, yielding a relatively poor strict preference relation.

In case the constant tradeoff hypothesis w.r.t. the g_i's is unrealistic, one can postulate like in the classical utility model, the existence of strictly monotonic rescalings u_i on each dimension. Letting $g'_i(a) = g_i(a) + q_i$, we could then consider the following aggregated model:

$$\begin{cases} a \succ b & \text{iff} \quad f(a) > f'(b), \\ a \sim b & \text{iff} \quad f(a) \leq f'(b) \text{ and } f(b) \leq f'(a), \end{cases}$$

$$\text{with} \begin{cases} f(a) & = \sum_{i=1}^{n} u_i\left(g_i(a)\right), \\ f'(a) & = \sum_{i=1}^{n} u_i\left(g_i'(a)\right). \end{cases} \qquad \textbf{(model M}_2\textbf{)}$$

In such a model, after rescaling by the u_i's, each alternative a is represented by an interval $[u_i(g_i(a)), u_i(g_i'(a))]$ on the ith axis and a is represented in the overall relation by the interval $[f(a), f'(a)]$. The overall preference is then the natural interval order on the intervals $\{[f(a), f'(a)], a \in A\}$ of the real line.

The latter model (M_2) boils down to (M_1) when the rescalings u_i simply consist in multiplying the g_i's by a constant factor w_i.

Generalizing the above models in a non-additive manner we get the following two new models yielding overall preference structures which are respectively semiorders and interval orders:

$$\begin{cases} a \succ b & \text{iff} \quad f(a) > f(b) + q, \\ a \sim b & \text{iff} \quad |f(a) - f(b)| \leq q, \end{cases}$$

$$\text{with} \begin{cases} f(a) & = f_g\left(g_1(a), \ldots, g_n(a)\right), \\ q & = q_g\left(q_1, \ldots, q_n\right), \end{cases} \qquad \textbf{(model M}_3\textbf{)}$$

f_g and q_g, two functions $\mathcal{R}^n \longrightarrow \mathcal{R}$, non decreasing in each argument, or

$$\begin{cases} a \succ b & \text{iff} \quad f(a) > f'(b), \\ a \sim b & \text{iff} \quad f(a) \leq f'(b) \text{ and } f(b) \leq f'(a), \end{cases}$$

$$\text{with} \begin{cases} f(a) = f_u\left(u_1(g_1(a)), \ldots, u_n(g_n(a))\right), \\ f'(a) = f_u'\left(u_1(g_1'(a)), \ldots, u_n(g_n'(a))\right), \\ u_i, i = 1, \ldots, n, \\ \text{strictly increasing functions: } \mathcal{R} \longrightarrow \mathcal{R}, \\ f_u, f_u', \text{ two functions: } \mathcal{R}^n \longrightarrow \mathcal{R}, \\ \text{non-decreasing in each of their arguments.} \end{cases} \qquad \textbf{(model M}_4\textbf{)}$$

6.5   Theoretical results related to the overall evaluation approach

There are very few results related to models M_3 and M_4 (and their particularizations M_1 and M_2); moreover they are mainly concerned with infinite sets of alternatives. We briefly describe some of the results obtained in Luce 1973, Gilboa and Lapson 1995; the interested reader is referred to those papers for more details and also to Suppes, Krantz, Luce and Tversky 1989, vol. II, section 16.6.

In both the above mentioned papers, the authors establish conditions under which a numerical representation of a semiorder on a product space (conjoint structure) can be obtained by combining numerical representations of "individual preferences". Let X be the cartesian product $\Pi_{i=1}^{n} X_i$ of the sets X_i and let $x = (x_1, \ldots, x_n)$ denote an element of X, with $x_i \in X_i$, $i = 1, \ldots, n$. A connection with models (M_3) and (M_4) can be established in case X is viewed as the set of vectors evaluating the alternatives $a \in A$, i.e.

$$X = \{(g_1(a), \ldots, g_n(a)), a \in A\}.$$

This presupposes that alternatives can be identified with their vector of evaluations (exhaustivity of the system of criteria) and that any combination of the evaluations on each criterion corresponds to a potential alternative. The latter is often a restrictive assumption since it may force the consideration of unrealistic alternatives in the construction of an overall preference relation $\succeq$ on X (e.g. a high performance car at low price).

Individual preference relations $\succeq_i$ may be defined on X_i, $i = 1, \ldots, n$, when an overall relation $\succeq$ is given on X.

Definition 6.2 *Let $\succeq$ be a binary relation on $X = \Pi_{i=1}^n X_i$. For each i, an individual preference relation $\succeq_i$ is defined on X_i; for $a_i, b_i \in X_i$:*

$$a_i \succeq_i b_i \text{ if } \exists x \in X \text{ s.t. } (a_i x_{-i}) \succeq (b_i x_{-i});$$

$(a_i x_{-i})$ *(resp. $(b_i x_{-i})$) is an obvious notation for the vector of X which has the same coordinates as $x \in X$ except for the ith coordinate which is a_i (resp. b_i).*

A much poorer individual preference relation would in general be obtained if "there exists" is substituted by "for all" in the above definition; both definitions are however equivalent in case the overall relation $\succeq$ satisfies the following independence axiom.

Definition 6.3 *(Independence). The overall relation $\succeq$ satisfies the independence property on all criteria if for all $i = 1, \ldots, n$, for all $a_i, b_i \in X_i$, for all $x, y \in X$,*

$$(a_i x_{-i}) \succeq (b_i x_{-i}) \text{ iff } (a_i y_{-i}) \succeq (b_i y_{-i}).$$

Theorem 6.9 *Let $\succeq$ be a binary relation on $X = \Pi_{i=1}^n X_i$. The relation $\succeq$ satisfies the independence property on all criteria*

$$\text{iff } [a_i \succeq b_i \text{ iff } \forall \, x \in X, \ (a_i x_{-i}) \succeq (b_i x_{-i})]$$

$$\text{iff } [a_i \succ b_i \text{ iff } \forall \, x \in X, \ (a_i x_{-i}) \succ (b_i x_{-i})].$$

The proof is immediate. Properties such as transitivity of $\succeq$ or the fact that $\succeq$ is a semiorder are automatically inherited by the individual relations $\succeq_i$.

Luce 1973 considers a semiorder $\succeq$ on a 2-factor product space $X = X_1 \times X_2$ and gives conditions under which a numerical representation f with constant threshold q can be obtained in an additive manner, i.e. $\exists \, g_1$ (resp. g_2) defined on X_1 (resp. X_2) such that

$$f(x_1, x_2) = g_1(x_1) + g_2(x_2)$$

and, $\forall \, x, y \in X$,

$$\begin{cases} x \succ y & \text{iff} \quad f(x) > f(y) + q, \\ x \sim y & \text{iff} \quad |f(x) - f(y)| \leq q. \end{cases}$$

In the class of tight representations, i.e. in case q is chosen such as

$$q = \max\left\{|f(x) - f(y)|,\ x, y \in X \text{ and } x \sim y\right\},$$

Luce's results also ensures unicity of the representation up to positive linear transformations. More precisely, if $f' = g_1' + g_2'$ is another tight representation with constant threshold q', then there are real constants $\alpha > 0$, β_1, β_2 such that

$$\begin{cases} g_i' &=& \alpha g_i + \beta_i \quad i = 1, 2, \\ q' &=& \alpha q. \end{cases}$$

Note that Luce's conditions involve "solvability" axioms which implicitly require sets X_i of infinite cardinality. Such axioms are needed for proving additivity of the overall representation as well as unicity up to positive linear transformations.

On finite sets, additive structures are usually characterized by means of "axiom schemes" (derived through the use of "theorems of the alternative") which are hardly interpretable in practical terms (Domotor and Stelzer 1971, Fishburn 1970). The fact that the overall relation is a semiorder instead of being a complete preorder does not change much to the difficulty of characterizing the additive structure.

Luce's results characterize in fact a particular case of model (M_1), in case X is infinite and is a product of 2 factors $X_1 \times X_2$. If the hypotheses are fulfilled, the individual preferences $\succeq_i$ are semiorders which can be represented by g_i, $i = 1, 2$, and the threshold q of the overall relation representation. So in this case, the weights w_i, $i = 1, 2$, in model (M_1) are both equal to $\frac{1}{2}$.

Gilboa and Lapson 1995 provide characterizations which come nearer to what is needed for preference modelling on a finite set of alternatives. We adapt here their results to the finite case.

Let $\succeq$ be a semiorder on $X = \Pi_{i=1}^n X_i$ which satisfies the independence axiom and suppose that the individual relations $\succeq_i$ are semiorders. Beside independence and a continuity axiom which is trivially satisfied for finite structures, they introduce a monotonicity property defined below.

Definition 6.4 *(Monotonicity) If T_i (resp. T) is the complete preorder associated with $\succeq_i$ (resp. $\succeq$) for all $i = 1, \dots, n$, $\succeq$ is said to be monotonic if*

$$[\forall\, i = 1, \dots, n,\ x_i T_i y_i] \Rightarrow x T y,$$

with $x = (x_1, \dots, x_n)$ and $y = (y_1, \dots, y_n) \in X$.

Let us call *regular* a numerical representation u of a semiorder S on X in which *equivalent* objects are represented by the same number. More precisely, denoting by E the equivalence relation associated with S, we have $x E y \Rightarrow u(x) = u(y)$.

Theorem 6.10 *(adapted from Gilboa and Lapson 1995, theorem A).*
Let $\succeq$ be a semiorder on $X = \Pi_{i=1}^n X_i$ with $|X_i| < \infty$, $i = 1, \dots, n$, and suppose that the individual relations $\succeq_i$ are semiorders on X_i, for all $i = 1, \dots, n$. Then, the following statements are equivalent:

(i) $\succeq$ *satisfies the independence and the monotonicity conditions;*

(ii) for all regular numerical representations u_i (resp. u) of $\succeq_i$ (resp. $\succeq$) with constant thresholds equal to 1, there exists a function $f_u : \mathcal{R}^n \longrightarrow \mathcal{R}$, strictly increasing in each variable, such that

$$u(x) = f_u(u_1(x_1), \ldots, u_n(x_n))$$

and the following equivalence holds for every $i \in \{1, \ldots, n\}$:

$$u_i(x_i) > u_i(y_i) + 1$$
$$\text{iff}$$
$$f_u\left(u_i(x_i), (u_j(z_j))_{j \in \{1, \ldots, n\}}^{j \neq i}\right) > f_u\left(u_i(y_i), (u_j(z_j))_{j \in \{1, \ldots, n\}}^{j \neq i}\right) + 1,$$
$$\forall x_i, y_i \in X_i, \forall z \in X.$$

The proof is given in section 6.9.

The above theorem implies that when aggregating semiordered individual preferences in a monotone and independent manner, overall preference is obtained as soon as there is individual preference on any criterion while the compared alternatives do not distinguish on the other criteria. In this model, overall preference is very sensitive to individual preference.

An example of an aggregation model which satisfies the hypothesis of Theorem 6.10 is obtained by defining $u(x) = \sum_{i=1}^{n} u_i(x_i)$. The difference with model (M_1) is that the overall threshold is not here a weighted sum (with positive weights) of the individual thresholds but rather the minimum of the individual thresholds. Model (M_3) encompasses the present case.

It is easy to generalize theorem 6.10 to interval orders and hence to obtain a particular case of model (M_4). The definition of monotonicity must be adapted.

Definition 6.5 *[Monotonicity IO]. If T_i^+ (resp. T^+) is the complete preorder associated with the left endpoints of the intervals $\succeq_i$ (resp. $\succeq$) and T_i^- (resp. T^-), the complete preorder associated with their right endpoints, the aggregation procedure $\succeq$ yielding an interval order is monotonic if*

$$\left[\forall i = 1, \ldots, n, \quad x_i T_i^+ y_i\right] \quad \Rightarrow \quad x T^+ y,$$

and

$$\left[\forall i = 1, \ldots, n, \quad x_i T_i^- y_i\right] \quad \Rightarrow \quad x T^- y.$$

Let us define a numerical representation of an interval order S on X as a pair of functions (u, u') defined on $X \longrightarrow \mathcal{R}$ with $x \succ y$ iff $u(x) > u'(y)$. If (u, u') is such a representation, $[u(x), u'(x)]$ can be seen as an interval representing x. The representation is *regular* if

$$x T^+ \cap \overline{T}^+ y \quad \Leftrightarrow \quad u(x) = u(y)$$

and

$$x T^- \cap \overline{T}^- y \quad \Leftrightarrow \quad u'(x) = u'(y).$$

Theorem 6.11 *Let $\succeq$ be an interval order on $X = \Pi_{i=1}^{n} X_i$ with $|X_i| < \infty$, $i = 1, \ldots, n$ and suppose that the individual relations $\succeq_i$ are interval orders on X_i, for all $i = 1, \ldots, n$. Then, the following statements are equivalent:*

(i) $\succeq$ satisfies the independence and the monotonicity conditions;

(ii) for all regular numerical representations (u_i, u_i') (resp. (u, u')) there exists a pair of functions (f_u, f_u'), each on $\mathcal{R}^n \longrightarrow \mathcal{R}$, strictly increasing in all n variables, such that

$$
\begin{aligned}
u(x) &= f_u(u_1(x_1), \ldots, u_n(x_n)), \\
u'(x) &= f_u'(u_1'(x_1), \ldots, u_n'(x_n))
\end{aligned}
$$

and the following equivalence holds for every $i \in \{1, \ldots, n\}$:

$$
u_i(x_i) > u_i'(y_i)
$$

$$
\text{iff}
$$

$$
f_u\left(u_i(x_i), (u_j(z_j))_{\substack{j \neq i \\ j \in \{1, \ldots, n\}}}\right) > f_u'\left(u_i'(y_i), (u_j'(z_j))_{\substack{j \neq i \\ j \in \{1, \ldots, n\}}}\right),
$$

$$
\forall\, x_i, y_i \in X_i, \forall\, z \in X.
$$

The proof of this theorem is in section 6.9.

Model (M_4) encompasses the situation described in the above theorem.

6.6   The pairwise comparisons paradigm

An alternative to the evaluation of all possible elements of A on an overall scale consists in comparing the actions in a pairwise manner and determining, for each pair of actions independently, to what extent an action is preferred to another. A number of procedures have been elaborated, based on various interpretations of this very vague scheme. The output of such procedures usually is a binary relation, possibly valued, on the set of alternatives.

Approaches of this type, based on the notion of "outranking" have been developped by B. Roy and gave rise to a series of methods (ELECTRE, PROME-THEE,...) for building an overall preference relation in a constructive manner. The concept of *outranking* is defined as follows (see Vincke 1992, p. 58, quoting Roy 1974):

> *An outranking relation is a binary relation S defined in A such that aSb if, given what is known about the decision maker's preferences and given the quality of the valuations of the alternatives and the nature of the problem, there are enough arguments to decide that a is at least as good as b, while there is no essential reason to refute that statement.*

The ELECTRE I method (Roy 1968) is the simplest example of an implementation of the outranking approach. Whether an alternative $a \in A$ is to be preferred to an alternative $b \in A$ in the overall relation is determined through a comparison of the "pros" and the "cons". Much like Condorcet's voting procedure

(Black 1958), a is globally preferred to b if the coalition of criteria on which a is not worse than b is "strong enough". This is tempered by possible "vetoes" of criteria against declaring that a is preferred to b. Such a veto occurs if b is so much better than a on some criterion that, in spite of the majority supporting a, it seems wise not to conclude. To be more specific, let a, b belong to the finite set A of alternatives; all alternatives are evaluated on n criteria through functions $g_i : A \longrightarrow \mathcal{R}$, $i = 1, \ldots, n$. Let $S(a, b)$ denote the set of criteria for which a is not worse than b, i.e. $\{i \in \{1, \ldots, n\} : g_i(a) \geq g_i(b) - q_i\}$ where q_i is a non-negative threshold called (individual) indifference threshold. Another threshold v_i, the veto threshold, must be defined for each criterion. For any pair of alternatives $a, b \in A$, a is said to "outrank" b and we note $a \succeq b$ iff

$$\frac{\sum_{i \in S(a,b)} w_i}{\sum_{i=1}^{n} w_i} \geq q,$$

and

$$\forall i = 1, \ldots, n, \quad g_i(a) > g_i(b) - v_i,$$

where w_i is a positive index associated with criterion i and reflecting its importance.

Such a particular approach can be situated in the perspective of general models based on pairwise comparisons of preference "differences". These models were investigated for taking into account the experimental evidence of lack of transitivity in overall preferences. The so-called "non transitive conjoint model" (Bouyssou 1986, Fishburn 1991b) provides a rather general framework. Let

$$a \succeq b \quad \text{iff} \quad \sum_i p_i \left(g_i(a), g_i(b)\right) \geq 0, \qquad \textbf{(model M$_5$)}$$

where p_i is a function $\mathcal{R}^2 \longrightarrow \mathcal{R}$, non decreasing in the first argument and non increasing in the second one, for $i = 1, \ldots, n$. Several particular choices for p_i are of interest, for instance, the case where the p_i's are skew-symmetric, i.e. $\forall x_i, y_i \in \mathcal{R}$,

$$p_i(x_i, y_i) = -p_i(y_i, x_i)$$

and the more general case

$$p_i(x_i, y_i) \cdot p_i(y_i, x_i) \leq 0.$$

A less general and earlier example of additive "measurement" without transitivity is the additive difference model discussed by Morrison 1962 and Tversky 1969. In this model, the overall preference is defined, for any pair a, b of alternatives by

$$a \succeq b \quad \text{iff} \quad \sum_{i=1}^{n} \phi_i \left(u_i(g_i(a)) - u_i(g_i(b))\right) \geq 0, \qquad \textbf{(model M$_6$)}$$

where, for all $i = 1, \ldots, n$, ϕ_i is a non-decreasing function : $\mathcal{R} \longrightarrow \mathcal{R}$ and u_i is a monotone function $\mathcal{R} \longrightarrow \mathcal{R}$. Note that in Tversky's model, the ϕ_i's are strictly increasing functions which we will not assume here since it would be inconvenient for dealing with semiordered individual preferences (see below in case the ϕ_i's are odd functions). Of course, model M_6 can be seen as particularizing M_5 by taking

$$p_i\left(g_i(a), g_i(b)\right) = \phi_i\left[u_i(g_i(a)) - u_i(g_i(b))\right],$$

i.e. p_i is a function of the difference of preference between a and b on criterion i (measured through a proper recoding u_i of the evaluations g_i). If the ϕ_i's are odd functions ($\phi_i(-x) = -\phi_i(x)$, $\forall x \in \mathcal{R}$), M_6 is then a skew symmetric M_5 model. In this case and if in addition the ϕ_i's are assumed to be strictly increasing functions for all $i = 1, \ldots, n$, we have

$$a \succeq_i b$$
$$\text{iff} \quad \phi_i\left[u_i(g_i(a)) - u_i(g_i(b))\right] \geq 0$$
$$\text{iff} \quad u_i(g_i(a)) - u_i(g_i(b)) \geq 0;$$

this means that the individual preference relations $\succeq_i$ are complete preorders, excluding general semiorders.

Although the intuition behind models M_5 and M_6 is rather different than for overall evaluation on a single criterion, it should be noted that M_5 and M_6 are general enough to encompass the ELECTRE I procedure as well as the additive models M_1 or M_2. In model M_2, of which M_1 is a particularization, we have

$$0 \leq f(a) - f(b) - q = \sum_i w_i\left[u_i(g_i(a)) - u_i(g_i(b)) - q_i\right].$$

It thus fits into model M_6 with

$$\begin{cases} \phi_i(x_i) = w_i(x_i - q_i), \\ x_i = u_i(g_i(a)) - u_i(g_i(b)). \end{cases}$$

For viewing ELECTRE I as a particular case of M_6, take

$$\phi_i(x_i) = \begin{cases} (1-q)w_i & \text{if} \quad x_i \geq -q_i, \\ -qw_i & \text{if} \quad -v_i < x_i < -q_i, \\ -M & \text{if} \quad x_i \leq -v_i, \end{cases}$$

with M, an arbitrarily large positive number, and u_i, the identity function on $\mathcal{R}$. Let $V(a,b)$ denote the set of criteria on which there is a veto of b against a and $T(a,b)$, the set of criteria

$$\{i \in \{1, \ldots, n\} \text{ s.t. } -v_i < g_i(a) - g_i(b) < -q_i\}.$$

We have

$$\sum_{i=1}^{n} \phi_i\left(u_i(g_i(a)) - u_i(g_i(b))\right) = \sum_{i \in S(a,b)} (1-q)w_i - q\sum_{i \in T(a,b)} w_i - M|V(a,b)|.$$

If M is large enough, this quantity will never be non-negative unless $V(a, b)$ is an empty set. If this is the case, the quantity will be non-negative iff

$$\sum_{i \in S(a,b)} w_i \geq q \sum_{i=1}^{n} w_i,$$

i.e.

$$\frac{\sum_{i \in S(a,b)} w_i}{\sum_{i=1}^{n} w_i} \geq q,$$

which corresponds to the ELECTRE I definition for deciding $a \succeq b$.

Figure 6.3 illustrates an essential difference between ELECTRE I and M_2. In M_2, the differences of preference are taken into account in a continuous manner since

$$f(a) - f'(b) = \sum_{i=1}^{n} [u_i(g_i(a)) - u_i(g_i'(b))],$$

with strictly increasing u_i's; in contrast, in ELECTRE I, the effect of preference differences is discretized by means of thresholds. This is to be put in relation with the initial motivation for introducing semiorders, i.e. the notion of "just noticeable difference".

6.7 A general framework

As usual for additive models, M_5 and M_6 can be characterized by means of the linear separation approach introduced by Kraft, Pratt and Seidenberg 1959, used by Scott 1964 and many others since then. When A is a finite set, this seems to be the only available approach and it is not very satisfactory as it does not allow for tests in practical situations. Fishburn 1991a provides a characterization of this type for binary relations $\succeq$ on a product space $X = X_1 \times X_2 \times \cdots \times X_n, n \geq 2, X$ finite, for which there is a numerical representation as in M_5 with skew-symmetric p_i's. In the same paper, other characterizations of the skew-symmetric version of M_5 are given in the case of infinite sets of alternatives.

Earlier, Bouyssou 1986 and Vind 1991 had considered the more general case without skew-symmetry. Bouyssou characterized the two-factor case ($n = 2$) while Vind gave a "topological" characterization valid for $n \geq 4$. Note that in the two-factor case, a characterization by simple axioms has been obtained but there is no hope for generalizing it to higher dimension (Bouyssou 1986). Since it seems difficult to avoid linear separation techniques in general characterizations of preference structures involving sums (such as M_5 or M_6) on finite sets, and since such characterizations are of little practical interest, we turn to more general models.

Goldstein 1991 introduces "decomposable threshold models" which allow to classify multidimensional stimuli in three categories ("good", "bad", "indifferent"). Interpreting the (oriented) pairs of alternatives as stimuli and the categories in terms of preferences, he gets the following model:

$$a \succeq b \text{ iff } F(p_1(g_1(a), g_1(b)), \ldots, p_n(g_n(a), g_n(b))) \geq 0, \quad \textbf{(Model M_7)}$$

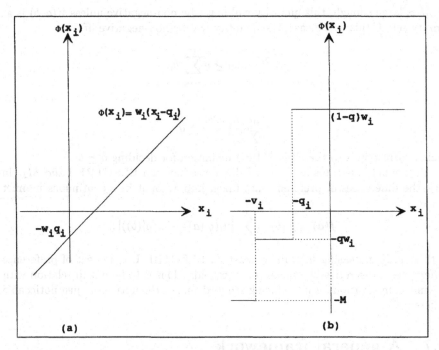

Figure 6.3: The function ϕ_i modelling the perception of preference differences on criterion i for M_2 (a) and ELECTRE I (b)

with $F : \mathcal{R}^n \longrightarrow \mathcal{R}$ and $p_i : \mathcal{R}^2 \longrightarrow \mathcal{R}$, $i = 1, \ldots, n$.

Conditions for the existence of such a representation are obtained in Goldstein 1991 under different hypotheses on F, including monotonicity and strict monotonicity in each of the n variables.

In the rest of this section, we follow Bouyssou and Pirlot 1997 and describe a class of binary relations $\succeq$ which satisfy (M_7) and provide a general framework for specifically dealing with the aggregation of semiorders (or interval orders). This framework is of particular interest in the context of the present work since the individual preferences $\succ_i$ induced by $\succeq$ are semiorders (or interval orders).

Let $X = \Pi_{i=1}^n X_i$ be a product of finite sets X_i (representing e.g. the values of the functions g_i on the set of alternatives A). We denote by X_{-i} the product of the $(n-1)$ sets $X_1 \times \cdots \times X_{i-1} \times X_{i+1} \times \cdots \times X_n$. Let $\succeq$ be a *reflexive* binary relation on X. A weak cancellation (WC) condition on $\succeq$ is first introduced in order to be able to represent $\succeq$ within (M_7).

(WC_i) For all $x_i, y_i, z_i, w_i \in X_i$ and for all $K, K', L, L' \in X_{-i}$,

$$
\left.
\begin{array}{ccc}
(x_iK) & \succeq & (y_iL) \\
 & \text{and} & \\
(z_iK') & \succeq & (w_iL')
\end{array}
\right\}
\Longrightarrow
\left\{
\begin{array}{ccc}
(x_iK') & \succeq & (y_iL') \\
 & \text{or} & \\
(z_iK) & \succeq & (w_iL).
\end{array}
\right.
$$

We say that $\succeq$ satisfies (WC) iff it satisfies (WC_i) for all $i = 1, \ldots, n$.
The (WC_i) property suggests that there is an ordering on the differences of preference on attribute i i.e. that either (x_i, y_i) is "larger" than (z_i, w_i) or the contrary. The following relation $\succeq_i^*$ defined on the pairs of values on each attribute i can be interpreted as comparing the differences of preference.

Definition 6.6 *For all* $x_i, y_i, z_i, w_i \in X_i$,

$$(x_i, y_i) \succeq_i^* (z_i, w_i)$$

iff for all $K, L \in X_{-i}$,

$$[(z_i, K) \succeq (w_i, L)] \Longrightarrow [(x_i, K) \succeq (y_i, L)].$$

The following theorems are given without the proofs which will appear in Bouyssou and Pirlot 1997.

Theorem 6.12 *For all* i, $\succeq_i$ *is transitive. The reflexive relation* $\succeq$ *on X satisfies* (WC_i) *for some* $i \in \{1, \ldots, n\}$ *iff* $\succeq_i^*$ *is complete; then* $\succeq_i^*$ *is a complete preorder.*

Theorem 6.13 *The reflexive relation* $\succeq$ *on X satisfies (WC) iff there is a function* $F : \mathcal{R}^n \longrightarrow \mathcal{R}$, *non-decreasing in each of the variables and functions* $p_i : \mathcal{R}^2 \longrightarrow \mathcal{R}$, $i = 1, \ldots, n$ *which represent the complete preorders* $\succeq_i^*$ *such that for all* $x = (x_1, \ldots, x_n)$, $y = (y_1, \ldots, y_n) \in X$,

$$x \succeq y \text{ iff } F[p_1(x_1, y_1), \ldots, p_n(x_n, y_n)] \geq 0.$$

We then introduce a second kind of cancellation properties (WC') which will ensure that the individual preference structures (in the sense of definition 6.2) are semiorders or interval orders. The (WC') cancellation property splits into three conditions for each criterion i, $(WC'1)_i$, $(WC'2)_i$ and $(WC'3)_i$.

$(WC_1')_i$ For all $x_i, y_i, z_i, w_i \in X_i$ and for all $K, K', L, L' \in X_{-i}$,

$$
\left.
\begin{array}{ccc}
(x_i, K) & \succeq & (y_i, L) \\
 & \text{and} & \\
(z_i, K') & \succeq & (w_i, L')
\end{array}
\right\}
\Longrightarrow
\left\{
\begin{array}{ccc}
(z_i, K) & \succeq & (y_i, L) \\
 & \text{or} & \\
(x_i, K') & \succeq & (w_i, L').
\end{array}
\right.
$$

$(WC_2')_i$ For all $x_i, y_i, z_i, w_i \in X_i$ and for all $K, K', L, L' \in X_{-i}$,

$$
\left.
\begin{array}{ccc}
(x_i, K) & \succeq & (y_i, L) \\
 & \text{and} & \\
(z_i, K') & \succeq & (w_i, L')
\end{array}
\right\}
\Longrightarrow
\left\{
\begin{array}{ccc}
(x_i, K) & \succeq & (w_i, L) \\
 & \text{or} & \\
(z_i, K') & \succeq & (y_i, L').
\end{array}
\right.
$$

$(WC_3')_i$ For all $x_i, y_i, z_i, w_i \in X_i$ and for all $K, K', L, L' \in X_{-i}$,

$$\left.\begin{array}{ccc}(x_i, K) & \succeq & (y_i, L) \\ & \text{and} & \\ (z_i, K') & \succeq & (x_i, L')\end{array}\right\} \Longrightarrow \left\{\begin{array}{ccc}(w_i, K) & \succeq & (y_i, L) \\ & \text{or} & \\ (z_i, K') & \succeq & (w_i, L').\end{array}\right.$$

We say that $\succeq$ satisfies (WC_1') (resp. (WC_2'), (WC_3')) iff it satisfies $(WC_1')_i$ (resp. $(WC_2')_i$, $(WC_3')_i$) for all $i = 1, \ldots, n$. Property $(WC_1')_i$ can be interpreted as telling that there is an ordering on the values taken by the alternatives on criterion i: x_i is either "larger" than z_i or the converse; $(WC_2')_i$ tells a similar thing. Comparing $(WC_1')_i$ to $(WC_2')_i$ suggests that the orderings on X_i may differ depending on whether the ith coordinate belongs to the description of an alternative which dominates another or is dominated by another. Taken together with $(WC_1')_i$ or $(WC_2')_i$, $(WC_3')_i$ implies that both orderings are identical. The following results are given without proofs which will appear in Bouyssou and Pirlot 1997.

Theorem 6.14 *If the reflexive relation $\succeq$ on X satisfies either $(WC_1')_i$ or $(WC_2')_i$, then the individual relation $\succeq_i$ on X_i is an interval order. If in addition $\succeq$ satisfies $(WC_3')_i$, then $\succeq_i$ is a semiorder.*

Theorem 6.15 *The reflexive relation $\succeq$ on X satisfies (WC), (WC_1'), and (WC_2') iff there are a function $F : \mathcal{R}^n \longrightarrow \mathcal{R}$, non decreasing in each coordinate and for, all $i = 1, \ldots, n$, functions $\psi_i : \mathcal{R}^2 \longrightarrow \mathcal{R}$, non decreasing in the first variable and non increasing in the second variable,*

$$u_i : X_i \longrightarrow \mathcal{R}$$
$$v_i : X_i \longrightarrow \mathcal{R}$$

such that for all $x, y \in X$,

$$F\left[(\psi_i(u_i(x_i), v_i(y_i)))_{i=1,\ldots,n}\right] \geq 0 \text{ iff } x \succeq y.$$

The relation $\succeq$ satisfies (WC_3') together with the other properties iff we have the above with $u_i \equiv v_i$.

With these results, we have a framework which is appropriate for dealing with the aggregation of semiorders (and also of interval orders); it generalizes models (M_5) and (M_6) and is precisely identified as a particularization of (M_7) (with special properties of F and the p_i's). This framework also contains models (M_1) to (M_4) as particular cases. Indeed, taking

$$x_i \text{ for } g_i(a), \ y_i \text{ for } g_i(b),$$
$$v_i(g_i(b)) = v_i(y_i) \text{ for } u_i(g_i'(b)),$$

and choosing the following particularizations of ψ_i and F :

$$\psi_i(u_i(x_i), v_i(y_i)) = u_i(x_i) - v_i(y_i),$$

$$F\left[(u_i(x_i) - v_i(y_i))_{i=1,\ldots,n}\right] = f_u(u_1(x_1), \ldots, u_n(x_n)) - f_u'(v_1(y_1), \ldots, v_n(y_n)),$$

we get

$$F\left[(u_i(x_i) - v_i(y_i))_{i=1,\ldots,n}\right] \geq 0$$
$$\text{iff} \quad f_u(u_1(x_1),\ldots,u_n(x_n)) \geq f'_u(v_1(y_1),\ldots,v_n(y_n))$$
$$\text{iff} \quad a \succeq b$$

as in model (M_4). Model (M_3) is obtained in a similar manner by taking further $u_i = v_i =$ identity and $f'_u(x_1,\ldots,x_n) = f_u(x_1,\ldots,x_n) - q$, $q \geq 0$.

So, the above framework encompasses both transitive and non-transitive preferences; it appears clearly that there is a range of procedures for aggregating criteria which essentially differ in the manner they deal with "preference differences"; these preference differences which are modelled by the relation $\succeq^*_i$ on the pairs $(x_i, y_i) \in X^2_i$, can be taken into account in a variety of methods ranging from model (M_2) with continuous ψ_i's (figure 6.3 a)) to ELECTRE I with discontinuous ψ_i's (figure 6.3 b)). Model (M_2) can be seen as a limit case of models with thresholds like ELECTRE I, for a number of thresholds becoming sufficiently large (a finite number of thresholds suffices in the case of a finite number of alternatives, but this number can grow without limit with the number of alternatives).

Another interesting question is how the present framework is connected to the theory of non-compensatory preference structures. The notion of non-compensation, first studied in Fishburn 1976 and generalized by Bouyssou and Vansnick 1986, has been used as a cornerstone for understanding the difference between the utility and "outranking" approaches, the latter being "nearly" non compensatory while the former is fully compensatory.

A binary preference relation $\succeq$ is non-compensatory if the fact that $a \succeq b$ only depends on the set of criteria on which a is better (or not worse) than b and the set of criteria on which b is better (or not worse) than a. In the framework introduced above one can describe such preference relations as corresponding to cases where the functions ψ_i very roughly deal with differences of evaluations, i.e. the ψ_i's take very few distinct values. A first example of a non-compensatory structure is provided by the output of the ELECTRE I method without veto threshold, i.e. the so-called *concordance* relation. This is obtained by the following choice of the ψ_i's which only take two distinct values according to the position of $(u_i(x_i) - u_i(y_i))$ w.r.t. a threshold:

$$\psi_i(u_i(x_i), u_i(y_i)) = \begin{cases} (1-q)w_i & \text{if } u_i(x_i) - u_i(y_i) \geq -q_i, \\ -qw_i & \text{otherwise,} \end{cases}$$

(with u_i, the identity function, in the original version of ELECTRE).

Another procedure of the same type but defined in terms of the asymmetric part of the individual preferences ("better" instead of "not worse") is the TACTIC method (Vansnick 1986) without veto threshold. This method was the essential motivation for Bouyssou and Vansnick 1986 generalizing the theory of non-compensatory structures. In TACTIC without veto threshold, we have

$$a \succ b \text{ iff } \sum_{i \in P(a,b)} w_i > q. \sum_{j \in P(b,a)} w_j,$$

where $P(a, b)$ (resp. $P(b, a)$) denotes the set of criteria on which a is strictly preferred to b (resp. b is strictly preferred to a). Such a procedure enters in the general framework with ψ_i's chosen as follows (see also the figure 6.4):

$$\psi_i(u_i(x_i), u_i(y_i)) = \left\{ \begin{array}{ll} w_i & \text{if} \quad u_i(x_i) - u_i(y_i) > q_i, \\ 0 & \text{if} \quad -q_i \leq u_i(x_i) - u_i(y_i) \leq q_i, \\ -qw_i & \text{if} \quad u_i(x_i) - u_i(y_i) < -q_i. \end{array} \right.$$

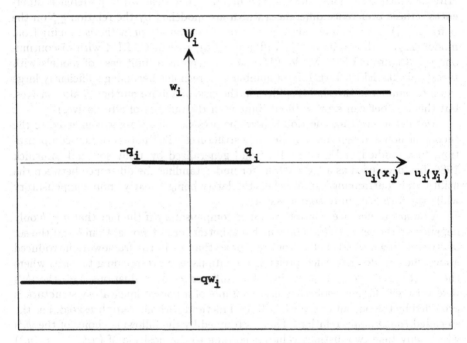

Figure 6.4: The function ψ_i for the TACTIC procedure

Although the opposition "compensatory vs. non-compensatory" is appealing, it follows from the work of Bouyssou and Vansnick 1986 that dealing with the valued versions of ELECTRE or even with the simpler versions with veto threshold seems hard in the context of generalized non-compensatory structures.

We feel that the new general framework presented in this section offers better opportunities for understanding the continuum of methods going from the purely non-compensatory to the fully compensatory ones. Much of this work remains to be done. In particular, the purely non-compensatory structures correspond to simple forms of ψ_i's and it could be of interest to characterize and study them.

6.8 Aggregation of valued semiorders and families of semiorders

In this section we consider another way of looking at the result of an aggregation procedure; this opens wide perspectives for theoretical investigation. Instead of viewing the output of the aggregation as a binary relation, one can alternatively consider that aggregation produces valued relations, the value attached to the pair of alternatives (a, b) being, in the most general model, the number

$$F\left[(\psi_i(u_i(x_i), u_i(y_i))_{i=1,\ldots,n}\right],$$

with $x_i = g_i(a)$, $y_i = g_i(b)$. This numerical index was exploited in a very elementary way in the previous sections; it was yielding a binary relation through checking whether it was non-negative ($F \geq 0$).

In the rest of this chapter, we consider aggregation procedures whose output is a valued relation on the set A of alternatives; we revisit already mentioned procedures like ELECTRE I as well as we introduce other procedures. We shall see also that most existing multicriteria aggregation methods fit into the framework of the aggregation of semiorders and more precisely of homogeneous families of semiorders.

Consider the following model, called (M_8), which is a direct generalization of the additive difference model (M_6) to the non-additive case; the output of the procedure is a valued relation.

Definition 6.7 *The valued relation* $s : A \times A \longrightarrow R$ *pertains to model* $(\mathbf{M_8})$ *if,*
$\forall\ a, b \in A,\ s(a, b) = F\left[(\phi_i(u_i(g_i(a)) - u_i(g_i(b)))_{i=1,\ldots,n}\right],$

where $F : \mathcal{R}^n \longrightarrow \mathcal{R}$ *is non-decreasing in each of its* n *variables,*
ϕ_i *and* $u_i : \mathcal{R} \longrightarrow \mathcal{R}$, *are non-decreasing for all* $i = 1, \ldots, n$.

Of course, it can be additionally imposed that the values of F range is some bounded interval, for instance, the $[0, 1]$ interval.

In such a model, for all $i = 1, \ldots, n$,

$$\Phi_i(a, b) = \phi_i\left[u_i(g_i(a)) - u_i(g_i(b))\right]$$

determines a valued semiorder on A; more precisely $\Phi_i(a, b)$ is a numerical representation of an homogeneous family $\succeq_{ij}$, $j = 1, \ldots, m_i$ of semiorders. If q_{ij} denote the m_i distinct values taken by the function Φ_i on A^2, ranked in increasing order (i.e. $q_{ij-1} \leq q_{ij}$, for all $j = 2, \ldots, m_i$), then the binary relations defined by

$$a \succeq_{ij} b \text{ iff } \phi_i\left(u_i(g_i(a)) - u_i(g_i(b))\right) = \Phi_i(a, b) \geq -q_{ij-1}$$

form an homogeneous family of semiorders.

In the sequel we illustrate the fact that most multicriteria aggregation procedures can be viewed as the aggregation of homogeneous families of semiorders into a valued relation (model M_8). It is the case with the additive utility model (M_2) where

$$s(a,b) = \sum_i [\phi_i(u_i(g_i(a))) - u_i(g_i(b))] = -s(b,a),$$

and of course with the more general model (M_3). It is also the case with the simple ELECTRE I method where $s(a,b)$ can be defined as

$$s(a,b) = \begin{cases} \frac{1}{w} \sum_{i \in S(a,b)} w_i & \text{if } g_i(a) - g_i(b) > -v_i, \quad \forall\, i = 1,\ldots,n, \\ -M & \text{if for some } i \text{ at least, } g_i(a) - g_i(b) \leq -v_i, \end{cases}$$

with $w = \sum_{i=1}^{n} w_i$ and M a large positive number. For all $i = 1, \ldots, n$, the homogeneous family of semiorders $\succ_{ij}$ of ELECTRE I consists at most of three semiorders $\succeq_{i1}, \succeq_{i2}, \succeq_{i3}$; for all $a, b \in A$,

$$\begin{array}{llll} a \succeq_{i1} b & \text{iff} & \Phi_i(a,b) \geq -M & \text{which is always true,} \\ a \succeq_{i2} b & \text{iff} & \Phi_i(a,b) \geq 0 & \text{iff } g_i(a) - g_i(b) > -v_i, \\ a \succeq_{i3} b & \text{iff} & \Phi_i(a,b) \geq w_i & \text{iff } g_i(a) - g_i(b) \geq -q_i. \end{array}$$

Hence we have

$$a \succeq_{i1} b \text{ and not } (a \succ_{i2} b) \text{ iff } g_i(a) - g_i(b) \leq -v_i$$

and one may take $\Phi_i(a,b) = -M$ in this case. Similarly,

$$a \succeq_{i2} b \text{ and not } (a \succ_{i3} b) \text{ iff } -v_i < g_i(a) - g_i(b) < -q_i$$

and one may take $\Phi_i(a,b) = 0$. In the case $a \succeq_{i3} b$, one takes $\Phi_i(a,b) = w_i$. With such values for Φ_i, F is simply the sum of its n variables unless at least one of them is equal to $-M$ in which case F itself is equal to $-M$. (For a related approach, see Pirlot 1996).

It can easily be seen that methods like ELECTRE III or IV (see Vincke 1992, pp. 64-69, for a description) or PROMETHEE pertain to model (M_8). Consider for instance the PROMETHEE II method (Brans and Vincke 1985). This method can be described as producing a valued relation defined for all pair $a, b \in A$ by

$$s(a,b) = \frac{1}{w} \sum_{i=1}^{n} w_i c_i(g_i(a) - g_i(b)),$$

where $w = \sum_{i=1}^{n} w_i$ and $\forall\, i$, $c_i : \mathcal{R} \longrightarrow \mathcal{R}$, is a function recoding the difference $((g_i(a) - g_i(b))$ according to one of the six types of behavior illustrated in figure 6.5.

It is interesting to notice that the six types of functions offer a good flexibility for modelling the differences of preference. Type VI offers a continuous strictly

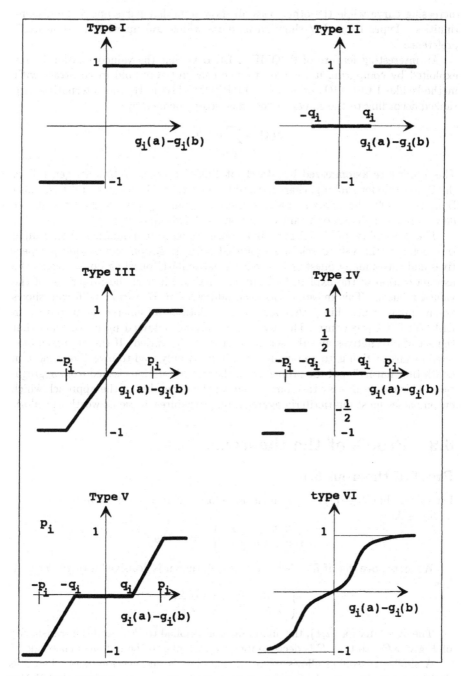

Figure 6.5: The six types of recoding of the difference of evaluations used in PROMETHEE II

increasing curve while the other ones all show some flat parts and some discontinuities. Types III and V show linear parts which end up with a "saturated" preference.

An interesting feature of PROMETHEE II is that the valued relation is not exploited by comparing it to zero or some constant threshold in constrast with methods like ELECTRE or utility. In PROMETHEE II, the alternatives are ranked according to the so-called "net flow score" defined by

$$\sigma(a) = \sum_{b \neq a} s(a,b).$$

This procedure axiomatized by Marchant 1996 for valued relations, generalizes the Borda rule for ranking, characterized by Nitzan and Rubinstein 1981 (see also Bouyssou 1992 who offers a characterization of ranking with respect to the net flow score on the basis of a valued relation on the alternatives).

The possibility of exploiting, in a more sophisticated manner than simple thresholding, the valued relation produced by aggregation opens wide perspectives and raises many questions as well. A particularly puzzling one concerns the axiomatization of the exploitation phase: what are indeed the properties of the valued relation. This question has been addressed in Bouyssou 1996 who shows among many other things that any valued relation can emerge as output of an ELECTRE III procedure. This result can only be obtained by postulating that the set of alternatives and the set of criteria can be varied. If the approach outlined in this section is to be pursued, it will inevitably lead to investigate more in depth the "nature" and the properties of the valued relations issued from aggregation. This is not an easy task but in view of the generality of the approach which encompasses most multicriteria aggregation procedures, it seems worth the effort.

6.9   Proofs of the theorems

Proof of theorem 6.1

Let us say that $E \subset \{1, 2, ..., n\}$ is decisive for (a, b) $(a \neq b)$ iff $\forall (S_1, S_2, ..., S_n) \in X$

$$\left. \begin{array}{l} a \, P_i \, b, \ \forall i \in E \\ b \, P_i \, a, \ \forall i \notin E \end{array} \right\} \Longrightarrow a \, P \, b.$$

We first prove that if E is decisive for (a, b), then it is decisive for (a, c), $\forall c \in A$. Let

$$\left\{ \begin{array}{l} a \, P_i \, b \, P_i \, c, \ \forall i \in E, \\ b \, P_i \, c \, P_i \, a, \ \forall i \notin E. \end{array} \right.$$

The decisivity for (a, b), the unanimity rule applied to (b, c) and the transitivity of P give aPc, so that E is decisive for (a, c), thanks to the independence rule.

A similar reasoning allows to prove that if E is decisive for (a, b), then it is decisive for (c, b), $\forall c \in A$. Combining these results, we can conclude that if E is decisive for (a, b) then it is decisive for every (c, d), so that we can call E a decisive set.

We now prove that if E is decisive and if $a\,P_i\,b, \forall i \in E$, then $a\,P\,b$. Let c be a third element of A and let

$$\begin{cases} a\,P_i\,c\,P_i\,b \ \forall i \in E, \\ c\,P_i\,a \text{ and } c\,P_i\,b, \ \forall i \notin E. \end{cases}$$

The decisivity of E (applied to (a,c)), the unanimity rule applied to (c,b) and the transitivity of P give aPb and the announced property follows from independence rule.

Let us prove that if F and G are disjoint and not decisive, then $F \cup G$ is not decisive. Let

$$\begin{cases} a\,P_i\,b\,P_i\,c\,P_i\,d, \ \forall i \in F, \\ c\,P_i\,d\,P_i\,a\,P_i\,b, \ \forall i \in G, \\ b\,P_i\,c\,P_i\,d\,P_i\,a, \ \forall i \notin F \cup G, \end{cases}$$

and assume $F \cup G$ is decisive.

The non-decisivity of F (applied to (a,d)) and the unanimity rule (applied to (c,d)) give $c\,P\,d\,(P \cup I)\,a$; with the decisivity of $F \cup G$ (applied to (a,b)) and the semiorder structure of S, this would imply cPb, which is in contradiction with the non-decisivity of G, so that $F \cup G$ cannot be decisive.

As the non-dictatorship rule implies that, $\forall i \in \{1, 2, ..., n\}, \{i\}$ is not decisive, we infer that $\{1\} \cup \{2\} \cup ... \cup \{n\} = \{1, 2, ..., n\}$ is not decisive, which is in contradiction with the unanimity rule.

Proof of theorem 6.2

Let $a_i\,P^{(1)}\,a_{i+1}$ for $i = 1, 2, ..., s$: $\forall i, \exists l_i$ such that $a_i\,P'_{l_i}\,a_{i+1}$ and $a_i\,I'_j\,a_{i+1}$, $\forall j < l_i$. Let $m = min\,\{\,l_i\,\}$; we have $a_i\,(P_m' \cup I_m')\,a_{i+1}, \forall i$, and $a_i\,P_m'\,a_{i+1}$ for at least one of them, so that $a_1\,P_m'\,a_{s+1}$; we have also, $\forall j < m, a_i\,I'_j\,a_{i+1}, \forall i$, hence $a_1\,I'_j\,a_{s+1}$. The conclusion is that we cannot have $a_{s+1}\,P^{(1)}\,a_1$.

Proof of theorem 6.3

Let $n = 2, a\,P^{(1)}\,b$ and $b\,P^{(1)}\,c$; this means that $a\,P_1\,b$ or $[a\,(P_1' \cup I_1')\,b$ and $a\,P_2\,b]$ and that $b\,P_1\,c$ or $[b\,(P_1' \cup I_1')\,c$ and $b\,P_2\,c]$. In three cases out of the four possible ones this gives $a\,P_1\,c$ and in the fourth case $[a\,(P_1' \cup I_1')\,c$ and $a\,P_2\,c]$, so that in all cases $a\,P^{(1)}\,c$.

Proof of theorem 6.6

Let $aP^{(3)}b$ and $b\,P^{(3)}\,c$; $\exists k_1$ and k_2 such that $aP_{k_1}b, a(P_j' \cup I_j')b, \forall j < k_1, bP_{k_2}c$ and $b(P_j' \cup I_j')c, \forall j < k_2$. Let $m = min\,\{k_1, k_2\}$; we have $a(P_j' \cup I_j')c, \forall j < m$, and either $aP_mb, b(P_m' \cup I_m')c$ or $a(P_m' \cup I_m')b, bP_mc$, so that aP_mc; it follows that $aP^{(3)}c$.

Proof of theorem 6.7

We will prove that there is a complete preorder (P', I') for which the matrix associated with $P^{(4)}$ is an upper diagonal step-type matrix. This complete preorder is obtained by the usual lexicographic rule applied to $\{ (P_1', I_1'), (P_2', I_2') \}$:

$$\begin{cases} a \, P' \, b \text{ iff } a \, P_1' \, b \text{ or } (a \, I_1' \, b \text{ and } a \, P_2' \, b), \\ a \, I' \, b \text{ iff } a \, I_1' \, b \text{ and } a \, I_2' \, b. \end{cases}$$

We have now to prove that

(a) $a \, (P' \cup I') \, b \implies b \, \neg P^{(4)} \, a$,

(b) $a \, P^{(4)} \, b$ and $x \, (P' \cup I') \, a \implies x \, P^{(4)} \, b$,

(c) $a \, P^{(4)} \, b$ and $b \, (P' \cup I') \, y \implies a \, P^{(4)} \, y$.

Proof of (a):

If $b \, P^{(4)} \, a$, then either $b \, P_1 \, a$ or $b \, P_1' \, a$ or $(b \, I_1' \, a$ and $b \, P_2 \, a)$; all these situations are incompatible with $a \, (P' \cup I') \, b$.

Proof of (b):

If $a \, P_1 \, b$ and $x \, (P' \cup I') \, a$, then $x \, (P_1' \cup I_1') \, a \, P_1 \, b$, which implies $x \, P_1 \, b$, hence $x \, P^{(4)} \, b$. If $a \, I_1 \, b, a \, P_1' \, b$ and $\exists c, d : a \, (P_1' \cup I_1') \, c \, I_1' \, d \, (P_1' \cup I_1') \, b$ and $c \, P_2 \, d$, then $x \, (P' \cup I') \, a$, which implies $x \, (P_1' \cup I_1') \, a$, gives $x \, P_1' \, b$, so that either $x \, P_1 \, b$ or $x \, I_1 \, b, x \, P_1' \, b$ and $\exists c, d : x \, (P_1' \cup I_1') \, c \, I_1' \, d \, (P_1' \cup I_1') \, b$ and $c \, P_2 \, d$. In both cases we obtain $x \, P^{(4)} \, b$. If $a \, I_1' \, b, a \, P_2 \, b$ and $x \, (P' \cup I') \, a$, then $x \, (P_1' \cup I_1') \, b$, leading to three cases:

(i) if $x \, P_1 \, b$ then $x \, P^{(4)} \, b$;

(ii) if $x \, I_1 \, b$ and $x \, P_1' \, b$, then $x \, P_1' \, a I_1' \, b$ and $a \, P_2 \, b$ imply $x \, P^{(4)} \, b$;

(iii) if $x \, I_1' \, b$ then $x \, I_1' \, a$; in this case $x \, (P' \cup I') \, a$ implies $x \, (P_2' \cup I_2') \, a$, which together with $a \, P_2 \, b$ gives $x \, P_2 \, b$, hence $x \, P^{(4)} \, b$.

Proof of (c): Similar to (b).

Proof of theorem 6.8

Substituting "P_2'" for "P_2" in the proof of theorem 6.7 yields a proof of our first assertion. As before we have $P^{(5)'} \subseteq P'$ and $I^{(5)'} \supseteq I'$. In addition, since both $a \, P_1' \, b$ and $(a I_1' \, b$ and $a \, P_2' \, b)$ imply $a \, P^{(5)'} \, b$, we get $a \, I^{(5)'} \, b$ iff $a \, (I_1' \cap I_2') \, b$, hence $I^{(5)'} = I'$ and $P^{(5)'} = P'$.

Proof of theorem 6.10

Let B denote the subset of $\mathcal{R}^n$ described by

$$B = \{(u_1(x_1), \ldots, u_n(x_n)), \; x = (x_1, \ldots, x_n) \in X\}.$$

1)$\Rightarrow$ 2). We first show that there is a *function* $f_u : B \longrightarrow \mathcal{R}$ such that, $\forall x \in X$,

$$f_u(u_1(x_1), \ldots, u_n(x_n)) = u(x).$$

Such a function exists iff, $\forall x, y \in X$,

$$[\forall i = 1, \ldots, n, \; u_i(x_i) = u_i(y_i)] \Rightarrow u(x) = u(y).$$

We have $u_i(x_i) = u_i(y_i)$ iff $x_i \, E_i \, y_i$ (with $E_i = T_i \cap \overline{T}_i$).
By monotonicity of $\succeq$ and using the regularity of the representation u, we get

$$[\forall i, x_i \, E_i \, y_i] \Rightarrow [\forall i, \; x_i \, T_i \, y_i \text{ and } y_i \, T_i \, x_i]$$

$$\Rightarrow x \, T \, y \text{ and } y \, T \, x \Rightarrow x \, E \, y \Rightarrow u(x) = u(y).$$

The function f_u is strictly increasing in each of its n variables since

$$u_i(x_i) > u_i(y_i) \Rightarrow x_i \, T_i \, y_i \text{ and not } (y_i \, T_i \, x_i).$$

So there must exist either $t_i \in X_i$ such that $x_i \succ_i t_i$ and not $(y_i \succ_i t_i)$, i.e. $t_i \succeq_i y_i$
or $s_i \in X_i$ such that $s_i \succ_i y_i$ and not $(s_i \succ_i x_i)$ i.e. $x_i \succeq_i s_i$. Consider the former
case; the latter being treated similarly. From the independence condition, we have
$\forall z \in X$,

$$f_u\left(u_i(x_i), (u_j(z_j))_{j \in \{1,\ldots,n\}}^{j \neq i}\right) > f_u\left(u_i(t_i), (u_j(z_j))_{j \in \{1,\ldots,n\}}^{j \neq i}\right) + 1$$

and

$$f_u\left(u_i(t_i), (u_j(z_j))_{j \in \{1,\ldots,n\}}^{j \neq i}\right) \geq f_u\left(u_i(y_i), (u_j(z_j))_{j \in \{1,\ldots,n\}}^{j \neq i}\right) - 1.$$

Hence $u(x_i, z_{-i}) > u(y_i, z_{-i})$.
From theorem 6.9 and the fact that $(u_i)_{i=1,\ldots,n}$ and u are representations with
threshold 1, we get that $\forall i, \; \forall x_i, y_i \in X_i$,

$$u_i(x_i) > u_i(y_i) + 1 \text{ iff } x_i \succ_i y_i,$$

whence $\forall z \in X$,

$$(x_i z_{-i}) \succ (y_i z_{-i}) \text{ iff } u(x_i z_{-i}) > u(y_i z_{-i}) + 1$$

iff

$$f_u\left(u_i(x_i), (u_j(z_j))_{j \in \{1,\ldots,n\}}^{j \neq i}\right) > f_u\left(u_i(y_i), (u_j(z_j))_{j \in \{1,\ldots,n\}}^{j \neq i}\right) + 1.$$

Finally, by "triangulating" the convex hull of B in $\mathcal{R}^n$ by means of hyper-tetrahedra (whose vertices belong to B and which do not contain any other point of B except for their vertices), it is possible to continue f_u into a function $\mathcal{R}^n \longrightarrow \mathcal{R}$

which is continuous, piecewise linear (linear on each tetrahedron) and strictly increasing. We still denote the function $\mathcal{R}^n \longrightarrow \mathcal{R}$ by f_u.

2) $\Rightarrow$ 1). Monotonicity of $\succeq$ follows immediately from strict monotonicity of f_u. For showing independence, let $x_i, y_i \in X_i$, $z, w \in X$; we have

$$
\begin{array}{rcl}
(x_i z_{-i}) & \succeq & (y_i z_{-i}) \\
\text{iff} \quad u(x_i z_{-i}) & \geq & u(y_i z_{-i}) - 1 \\
\text{iff} \quad u(y_i z_{-i}) & \leq & u(x_i z_{-i}) + 1 \\
\text{iff} \quad u_i(y_i) & \leq & u_i(x_i) + 1 \\
\text{iff} \quad u(y_i w_{-i}) & \leq & u(x_i w_{-i}) + 1 \\
\text{iff} \quad (x_i w_{-i}) & \succeq & (y_i w_{-i}).
\end{array}
$$

Proof of theorem 6.11

Let B, B' denote respectively the subsets of $\mathcal{R}^n$ obtained as image of X by the vectors of functions $(u_1, \ldots, u_n)$ and $(u'_1, \ldots, u'_n)$.

1) $\Rightarrow$ 2). Using the regularity of the representations and the monotonicity of $\succeq$, we get the existence of functions f_u on B and f'_u on B' exactly as in the proof of theorem 6.10. We now show that f_u is strictly increasing on B. Let $x_i, y_i \in X_i$ be such that $u_i(x_i) > u_i(y_i)$. Due to the postulated regularity of (u_i, u'_i), this implies that there is $t_i \in X_i$ such that $x_i \succ_i t_i$ and not $(y_i \succ_i t_i)$. This yields

$$
f_u\left(u_i(x_i), (u_j(z_j))_{\substack{j \neq i \\ j \in \{1, \ldots, n\}}} \right) > f'_u\left(u'_i(t_i), (u'_j(z_j))_{\substack{j \neq i \\ j \in \{1, \ldots, n\}}} \right)
$$

and

$$
f_u\left(u_i(y_i), (u_j(z_j))_{\substack{j \neq i \\ j \in \{1, \ldots, n\}}} \right) \leq f'_u\left(u'_i(t_i), (u'_j(z_j))_{\substack{j \neq i \\ j \in \{1, \ldots, n\}}} \right).
$$

Hence

$$
f_u\left(u_i(x_i), (u_j(z_j))_{\substack{j \neq i \\ j \in \{1, \ldots, n\}}} \right) > f_u\left(u_i(y_i), (u_j(z_j))_{\substack{j \neq i \\ j \in \{1, \ldots, n\}}} \right).
$$

A similar reasoning leads to the proof that f'_u is strictly increasing on B'.

The last property of f_u and f'_u is obtained in the following manner. Suppose $x_i, y_i \in X_i$ and $x_i \succ_i y_i$, which means that $u_i(x_i) > u'_i(y_i)$. By theorem 6.10, $\forall z \in X$,

$$
\begin{array}{rcl}
(x_i z_{-i}) & \succ & (y_i z_{-i}) \\
\text{iff} \quad u(x_i z_{-i}) & > & u'(y_i z_{-i})
\end{array}
$$

$$
\text{iff} \quad f_u\left(u_i(x_i), (u_j(z_j))_{\substack{j \neq i \\ j \in \{1, \ldots, n\}}} \right) > f'_u\left(u'_i(y_i), (u'_j(z_j))_{\substack{j \neq i \\ j \in \{1, \ldots, n\}}} \right).
$$

Finally, both f_u and f'_u can be continued in order to be defined on $\mathcal{R}^n$ (see the proof of theorem 6.10.)

2) $\Rightarrow$ 1). Monotonicity of $\succ$ w.r.t. T_i^+ (resp. T_i^-) follows from the strictly increasing character of f_u (resp. f'_u). For showing independence of $\succeq$, let $x_i, y_i \in$

X_i, $z, w \in X$; we have

$$
\begin{array}{rccc}
\text{iff} & (x_i z_{-i}) & \succeq & (y_i z_{-i}) \\
\text{iff} & u(y_i z_{-i}) & \leq & u'(x_i z_{-i}) \\
\text{iff} & u(y_i) & \leq & u'(x_i) \\
\text{iff} & u(y_i w_{-i}) & \leq & u'(x_i w_{-i}) \\
\text{iff} & (x_i w_{-i}) & \succeq & (y_i w_{-i}).
\end{array}
$$

7

MISCELLANEOUS

In this chapter, we mention briefly some other subjects in relation with semiorders, which were treated in the literature but will not be developed in this book.

7.1   Infinite semiorders

All the results presented in this book are concerned with semiorders on finite sets; this section gives some indications for the infinite case.

First of all, it is clear that the numerical representation of a semiorder with a constant threshold is not always possible in the infinite case. However, in the countable case, the conditions we have used to define the structure of semiorder are still necessary and sufficient for a numerical representation with a (variable) threshold.

Theorem 7.1 *Given a relation $S = (P, I)$ on a countable set A, the necessary and sufficient condition for the existence of two real valued functions g and q, defined on A, such that, $\forall\, a, b \in A$,*

$$
\left\{
\begin{array}{ll}
aPb & \Leftrightarrow \quad g(a) > g(b) + q(b), \\
aIb & \Leftrightarrow \quad \left\{
\begin{array}{l}
g(a) \leq g(b) + q(b), \\
g(b) \leq g(a) + q(a),
\end{array}
\right. \\
q(a) & \geq 0,
\end{array}
\right.
$$

is that, $\forall\, a, b, c, d \in A$,

$$
\left\{
\begin{array}{llll}
aPb, & bIc, & cPd & \Rightarrow \quad aPd, \\
aPb, & bPc, & cId & \Rightarrow \quad aPd.
\end{array}
\right.
$$

In the uncountable case, it is necessary to assume the existence in A of a "dense countable subset" in order to be able to have a threshold-representation of a semiorder. For more details and proofs, the interested reader is referred to Vincke 1978, Fishburn 1985, Suppes et al. 1989.

7.2   Semiordered mixture sets

Much of utility theory is applied to situations involving decision making under risk. A classic set of axioms is that of Von Neumann and Morgenstern 1947, which insures the existence of linear utility function on a completely preordered mixture set (see also Herstein and Milnor 1953). The question of the generalization of this set of axioms to semiordered mixture sets was set by Luce 1956. An answer was given by Vincke 1980 and used for preference representation in actuarial sciences (Vincke 1979).

7.3   Partial semiorders and interval orders

In terms of preference representation, a semiorder (or an interval order) can be seen as composed of a strict preference relation (asymmetric part of the relation, and an indifference relation (symmetric part). Every absence of strict preference between two alternatives is thus considered as an indifference. However, it often happens, in decision problems, that an absence of preference is due to either lack of information or conflictual information (as, for example, in a multicriteria or a multiactor context). In both cases, it is more realist to consider an absence of preference between two alternatives as an incomparability between them.

This is a motivation for studying triplets (P, I, J) where P is a strict preference relation (asymmetric), I is an indifference relation (reflexive and symmetric) and J is an incomparability relation (irreflexive and symmetric). In this context, propositions have been made in the litterature for defining *partial semiorders* or *partial interval orders* which, beyond the phenomenon of indifference threshold, express the possible presence of incomparability. We recall here, without proofs, some results of Roubens and Vincke 1985. Other references on this subject are Doignon et al. 1986 and Roubens and Vincke 1984.

Definition 7.1 *The triplet (P, I, J) is a partial semiorder on the finite set A iff P is transitive and every circuit of $P \cup I$ contains more I than P.*

Definition 7.2 *The triplet (P, I, J) is a partial interval order on the finite set A iff P is transitive and relation PI contains no circuit.*

Theorem 7.2 *If (P, I, J) is a partial semiorder on the finite set A, then there exists a real valued function g, defined on A, and a nonnegative constant q such that, $\forall\, a, b \in A$,*

$$\left\{ \begin{array}{lll} aPb & \Rightarrow & g(a) > g(b) + q, \\ aIb & \Rightarrow & |g(a) - g(b)| \le q. \end{array} \right.$$

Theorem 7.3 *If (P, I, J) is a partial semiorder on the finite set A, then there exist two real valued functions g and q, defined on A, such that, $\forall\, a, b \in A$,*

$$\left\{ \begin{array}{lll} aPb & \Rightarrow & g(a) > g(b) + q(b), \\ aIb & \Rightarrow & \left\{ \begin{array}{l} g(a) \le g(b) + q(b), \\ g(b) \le g(a) + q(a), \end{array} \right. \\ q(a) & \ge 0, \end{array} \right.$$

Theorem 7.4 *If (P, I, J) is a partial semiorder (resp. interval order) on the finite set A, then there exists a semiorder (resp. interval order) (P', I') on A such that $P \subseteq P'$ and $I \subseteq I'$.*

7.4 Double threshold models

Several multicriteria decision-aid methods (ELECTRE III and IV, PROMETHEE) assume that the preferences of the decision-maker, according to each criterion, can be represented by a real valued function g and two thresholds q and p such that, $\forall \, a, b \in A$,

$$
\left\{
\begin{array}{lll}
aPb & \Leftrightarrow & g(a) > g(b) + p(b), \\
aQb & \Leftrightarrow & g(b) + p(b) \geq g(a) > g(b) + q(b), \\
aIb & \Rightarrow & \left\{ \begin{array}{l} g(b) + q(b) \geq g(a), \\ g(a) + q(a) \geq g(b), \end{array} \right. \\
0 & \leq q(a) \leq & p(a),
\end{array}
\right.
$$

where Q is an asymmetric relation (called *weak preference* by Roy 1985) representing a situation of hesitation between indifference (I) and strict preference (P). Of course, this model can be seen as a particular case of a valued semiordered relation (studied in chapter 5), but it was specifically studied relatively to seven possible "coherence conditions" on the thresholds, like for example,

$$ ga) + q(a) > g(b) + q(b) \Leftrightarrow g(a) + p(a) > g(b) + p(b), $$

or

$$ g(a) > g(b) + q(b) \Rightarrow g(a) + q(a) > g(b) + p(b). $$

As proved in Vincke 1988, the necessary and sufficient condition for representing the triplet (P, Q, I) of relations by the above numerical model is that, $\forall \, a, b, c, d \in A$,

$$
\left\{
\begin{array}{llll}
aQb, & bIc, & cQd & \Rightarrow \quad a(Q \cup P)d, \\
aQb, & bIc, & cPd & \Rightarrow \quad aPd, \\
aPb, & bIc, & cPd & \Rightarrow \quad aPd, \\
aPb, & b\overline{Q}c, & cPd & \Rightarrow \quad aPd,
\end{array}
\right.
$$

In the same paper, necessary and sufficient conditions are given to obtain the above numerical representation together with one or several "coherence conditions". As a particular case, a characterization is provided of the so-called *pseudo-orders*, presented by Roy and Vincke 1987.

7.5 Dimension of a strict semiorder

A strict semiorder is a transitive and asymmetric relation. It is well-known that such a relation can be seen as the intersection of several strict complete orders and that the *dimension* is the minimum number of strict complete orders whose intersection provides the given relation (Dushnik and Miller 1941).

It is interesting to stress the fact that the dimension of a strict semiorder is always less than or equal to 3 (Rabinovitch 1978), while no finite bound exists for strict interval orders (Bogart, Rabinovitch and Trotter Jr. 1976).

A consequence is that the dimension of strict semiorders can be computed in polynomial time (because it is possible to determine in polynomial time whether or not a given order has dimension 2). Nothing is known about the complexity status for the dimension problem of strict interval orders and for counting the linear extensions of a strict semiorder or a strict interval order.

In connection with other notions of dimension, note also that several variants of the *semiorder dimension* of a relation have been defined in the literature; these are related to the minimal number of semiorders whose intersection is a given relation (Doignon, Ducamp and Falmagne 1987, Flament 1983).

7.6   Enumerating semiorders

The set of semiorders on a finite set is obviously much richer than the set of complete orders or preorders. As an example, the number of semiorders on a set of 7 elements is 763,099 while the number of complete preorders is 47,293 and the number of complete orders is 5,040. Chandon, Lemaire and Pouget 1978 have established a recurrence relation allowing to calculate the number of semiorders on a set of n elements, as a function of n. If we consider as equivalent, semiorders which are identical up to a permutation of the elements, it can be proved that the number of non equivalent semiorders on a set of n elements is given by $\frac{1}{n+1}\binom{2n}{n}$, the n-th *Catalan number* (see also Mitas 1994).

The interested reader is invited to consult Wine and Freund 1957, Dean and Keller 1968, Sharp 1971/72, Rogers 1977, Chandon et al. 1978.

7.7   Computational aspects

A first group of algorithmic questions is related with problems such as recognition, computation of a representation or of a characteristic. Recognizing that a binary relation is a semiorder or not and finding the minimal representation can both be done in linear time (and space) w.r.t. the number of nodes plus the number of arcs (Mitas 1994). An interesting question also raised in Mitas 1994 is to identify the reasons why several problems are known to be polynomial for semiorders while being NP-hard or of unknown complexity status for interval orders.

Another class of interesting problems can be described as approximation problems as for instance approximating a relation by a semiorder or approximating a semiorder by a preorder. Little is known on this subject. The problem of finding a semiorder at minimal distance (symmetric difference) to a binary relation is NP-hard (Hudry 1989) and so is the *bandwidth* problem (see Garey and Johnson 1979, p.200, for the "decision" version) , a similar problem with another objective function. Note that a number of approximation problems for interval orders and interval graphs have been proved to be NP-hard (see e.g. Golumbic and Shamir 1993).

Since the approximation problems seem to be hard, heuristics can be considered. In (Jacquet–Lagrèze 1978), an algorithm is proposed for finding approximations of a binary relation by various semiorders according to several criteria. The obtained semiorders are more or less faithful (with respect to the symmetric and the asymmetric parts of the original relation as well as to the associated incomparability relation (see section 7.3) and more or less discriminating (i.e. with large asymmetric or large symmetric part). An algorithm is also proposed in the same paper, for approximating probabilistic valued relations ($v(a, b) + v(b, a) = 1$, see chapter 5.2) by linear semiordered valued relations.

7.8 Indifference graphs and families of indifference graphs

At several places in this book, we have alluded to situations where the asymmetric part of a semiorder is not the important one. It is in particular the case in classification (see chapter 2, section 11) where the symmetric part of the semiorders, the *indifference graphs* are studied for themselves. Indifference graphs are also of interest as a particular class of perfect graphs. Let us mention that some effort has recently been devoted to the study of very general ordered structures whose symmetric part belongs to special classes of co-comparability graphs or more general perfect graphs (Abbas and Vincke 1993, Abbas 1994, Abbas, Pirlot and Vincke 1996, Fishburn 1997).

We collect below (and state more formally) a number of equivalent definitions and properties of indifference graphs, most of them having been evoked elsewhere in this book.

Theorem 7.5 *Let (A, I) be a non-directed graph on the set A of nodes (with a loop at each node). The following properties are equivalent.*
(i) There exist a function g and a non-negative function $q : A \longrightarrow \mathcal{R}$, such that $\forall\, a, b \in A, \, a \neq b$,

$$\{a, b\} \in I \quad \textit{iff} \quad \left\{ \begin{array}{l} g(a) \leq g(b) + q(b) \\ g(b) \leq g(a) + q(a) \end{array} \right.$$

and

$$g(a) > g(b) \quad \Rightarrow \quad g(a) + q(a) > g(b) + q(b).$$

(ii) There exist a function $g : A \longrightarrow \mathcal{R}$ and a number $q \geq 0$ such that, for all $a, b \in A, \, a \neq b$,

$$\{a, b\} \in I \textit{ iff } |g(a) - g(b)| \leq q.$$

(iii) The symmetric adjacency matrix $(I_{ij}, \, i, j \in A)$ where $I_{ij} = 1$ if $i = j$ or $\{i, j\} \in I$ and 0 otherwise, has the consecutive 1's property, i.e. it is possible to order the nodes of A in such a way that there never is a 0 between 1's in any row of the matrix (nor in any column since the matrix is symmetric).

Proof of the theorem

The first two conditions imply the existence of a semiorder represented by g and q, whose symmetric part is precisely the indifference graph. The two conditions are equivalent according to theorem 3.1. The third condition implies that the matrix S obtained by filling matrix (I_{ij}, $i,j \in A$) with 1's all above the main diagonal is the step-type matrix of a semiorder whose symmetric part is I and this definition is equivalent to the first two, as observed in chapter 3, section 17. A direct proof of this is in Roberts 1968.

Indifference graphs are a particular case of interval graphs describing the intersections of sets of intervals of the real line. Indifference graphs are sometimes called *unit interval graphs* since they can be represented by intervals of unit length as is apparent from property (ii) in the above theorem (with $q = 1$). They are also called *proper interval graphs*, in view of property (i), because they can be represented by intervals which are not properly included in one another. Note that a characterization of indifference graphs by means of forbidden induced subgraphs is also known (see Roberts 1969).

The link between indifference graphs and semiorders has been made clear in Roberts 1971b. The correspondance is essentially one-to-two; any indifference graph is the symmetric part of exactly two semiorders which are the converse of each other, provided the indifference graph is connected and reduced ("reduced" means that there are no equivalent nodes, i.e. no two nodes are linked to exactly the same subset of nodes).

A transition to the "valued" case, i.e. to (homogeneous) families of indifference graphs, is provided by the so-called *robinsonian* matrices. The robinsonian property, introduced in Kendall 1969a and named after Robinson (Robinson 1951), is a natural generalization of the consecutive 1's property. A symmetric matrix is *robinsonian* if all values on the main diagonal are equal and the entries in each row do not decrease as the main diagonal is approached. Such matrices are interesting cases of *similarity* matrices. Without loss of generality, we may impose that all entries are between 0 and 1 and even that the minimal one is 0 while the values on the main diagonal are all 1 (this can be obtained by eventually applying a positive affine transformation to the original values). If we substitute all above-diagonal elements by 1's, we obtain the matrix of a linear semiordered valued relation; the λ-cuts of this relation constitute an homogeneous family of semiorders whose symmetric parts can be obtained as λ-cuts of the original robinsonian matrix. This results immediately from theorem 5.8 (see also Roberts 1978). Notice that the robinsonian *dissimilarity* matrices considered in section 11 of chapter 2 can be viewed as the asymmetric part of the matrix of a linear semiordered valued relation. For more details on valued indifference graphs as well as on their applications in classification and seriation, the reader is referred to Kendall 1969a, Roberts 1971b, Hubert 1974, Roberts 1978, Bertrand 1992.

CONCLUSION

In writing this book we wanted to put together a number of concepts and results which could be useful for supporting the development of practical methods and tools in view of applications. What are the main ideas on which we put a stress in this book, that is what we briefly want to summarize as a conclusion.

In chapter 4, 5 and 6, we have studied semiorders (on finite sets), as extensively as we could, along three main axes. In chapter 4, the finite structure of semiorders was investigated; this yielded results related with special numerical representations which can be interpreted as generalizing the rank as defined for complete orders; those investigations also reveal a basic structure which summarizes a semiorder, namely its super synthetic graph. In developing this, we wanted to make semiorders almost as familiar and intuitive as ordinary complete orders, for which it is quite natural to use the rank for labelling the ordered objects. With semiorders we are somewhere between pure ordinality and cardinality or values. Essentially, this is this fact that is further illustrated in the next two chapters.

The important idea underlying chapter 5 is that ordinally-valued relations are equivalent to families of binary relations (corresponding to cuts at different value levels). By ordinally-valued, we mean that the values are defined up to a strictly increasing transformation. Under some simple assumptions, the cuts are semiorders and these semiorders form a more or less coherent family depending on additional properties of the valued relation. In the model with the strongest structure, all semiorders of the family can be represented by a single function and various thresholds. One step beyond, on the way from pure ordinality to cardinality is accomplished here since it is shown that families of semiorders can arise from looking at the objects or comparisons of objects with various levels of resolution or coarse-graining; quantities or their differences can be categorized according to a number of levels which reflect pertinent degrees of discrepancy in a given context.

In contrast with the previous situation where the information on the object is generally coherent and detailed (since objects are compared at several levels of discrepancy), we address in chapter 6, the problem of aggregating contradictory or conflicting informations. Besides showing the difficulty of getting regular and interpretable structures from the aggregation of semiorders, we describe a very general framework which appears to be well-suited for aggregation of semiordered information. Many of the procedures currently used for aggregating preferences fit into this scheme. Again in this model, the most important is the way in which differences of preference are taken into account in the aggregation phase.

So, what we actually wanted to show is that semiorders (or families of them) constitute an intermediate level between ordinal information and value. There are several contexts, in particular in preference modelling, where one has the feeling that numbers do not really mean what they seem to tell us but still more than simply their ordering; semiorders may constitute a powerful intermediate model. This may be the case not only in preference modelling and aggregation but more generally in the modelling of imprecision. Intervals of imprecision around a measure (obtained e.g. through using a measurement apparatus) might be associated degrees of credibility and doing so might be more relevant, from an interpretative viewpoint, than a statistical theory, especially when the measurement cannot be repeated.

Although there are a number of fields in which semiordered-based methods are already used or could easily be used (see chapter 2), we believe that there is still some effort to be done in order to operationalize the concepts we have tried to illustrate in this book. These results and concepts are in a sense halfway between theory and practice. We hope we have gathered some arguments to convince both practitioners and researchers to get interested in semiorders and related order structures such as interval orders. Major efforts should now be devoted to the design of practical methods for eliciting ordered structures and dealing efficiently with them in view of some particular goals (construction of a global preference, classification, transformation of imprecise input data using a model, etc). While doing this, there is no doubt that new and interesting theoretical problems will arise and we hope that the material gathered here could be of some use to help to solve them.

Bibliography

[1] Abbas, M. (1994). *Contribution au rapprochement de la théorie des graphes et de l'aide à la deécision: graphes parfaits et modèles de préférence*, PhD thesis, Université Libre de Bruxelles, Belgium.

[2] Abbas, M., Pirlot, M. and Vincke, Ph. (1996). Preference structures and co-comparability graphs, *Journal of Multicriteria Decision Analysis* **5**: 81–98.

[3] Abbas, M. and Vincke, Ph. (1993). Preference structures and threshold models, *Journal of Multicriteria Decision Analysis* **2**: 171–178.

[4] Ahuja, R.K., Magnanti, T.L. and Orlin, J.B. (1993). *Network flows. Theory, Algorithms and Applications*, Prentice–Hall.

[5] Armstrong, W.E. (1939). The determinateness of the utility function, *Economics Journal* **49**: 453–467.

[6] Arrow, K. (1963). *Social choice and individual values*, 2nd edn, Wiley.

[7] Bana e Costa, C. and Vansnick, J.-C. (1994). Macbeth – an interactive path towards the construction of cardinal value functions, *International Transactions in Operational Research* **1**(4): 489–500.

[8] Bell, D.E., Raiffa, H. and Tversky, A. (1988). *Decision making: descriptive, normative and prescriptive interactions*, Cambridge University Press.

[9] Benzer, S. (1962). The fine structure of the gene, *Scientific American* **206**(1): 70–84.

[10] Bertrand, P. (1992). Propriétés et caractérisations topologiques d'une représentation pyramidale, *Mathématiques et Sciences Humaines* **117**: 5–28.

[11] Black, D. (1958). *The theory of committees and elections*, Cambridge University Press.

[12] Bogart, K., Rabinovitch, I. and Trotter Jr., W. (1976). A bound on the dimension of interval orders, *Journal of Combinatorial Theory (A)* **21**: 319–328.

[13] Bouyssou, D. (1986). Some remarks on the notion of compensation in MCDM, *European Journal of Operational Research* **26**: 150–160.

[14] Bouyssou, D. (1992). Ranking methods based on valued preference relations: a characterization of the net flow method, *European Journal of Operational Research* **60**: 61–67.

[15] Bouyssou, D. (1996). Outranking relations: do they have special properties, *Journal of Multicriteria Decision Analysis* **5**: 99–111.

[16] Bouyssou, D. and Pirlot, M. (1997). A general framework for the aggregation of semiorders, *Technical report*, ESSEC, Cergy-Pontoise.

[17] Bouyssou, D. and Vansnick, J.-C. (1986). Noncompensatory and generalized noncompensatory preference structures, *Theory and Decision* **21**: 251–266.

[18] Bradley, S.P., Hax, A.C. and Magnanti, T.L. (1977). *Applied Mathematical Programming*, Addison-Welsey.

[19] Brans, J. P. and Vincke, Ph. (1985). A preference ranking organization method, *Management Science* **31**(6): 647–656.

[20] Carrano, A.V. (1988). Establishing the order of human chromosome-specific dna fragments, *in* A. Woodhead and B. Barnhart (eds), *Biotechnology and the Human Genome*, Plenum Press, 37–50.

[21] Chandon, J., Lemaire, J. and Pouget, J. (1978). Dénombrement des quasi-ordres sur un ensemble fini, *Mathématiques et Sciences Humaines* **62**: 61–80.

[22] Coombs, C.H. and Smith, J.E.K. (1973). On the detection of structures in attitudes and developmental processes, *Psychological Review* **80**: 337–351.

[23] Cozzens, M.B. and Roberts, F.S. (1982). Double semiorders and double indifference graphs, *SIAM Journal on Algebraic and Discrete Methods* **3**: 566–583.

[24] Cuthill, E. and McKee, J. (1969). Reducing the bandwidth of sparse symmetric matrices, *24th National Conference of ACM*, 157–172.

[25] Dean, R. and Keller, G. (1968). Natural partial orders, *Canadian Journal of Mathematics* **20**: 535–554.

[26] Debreu, G. (1960). Topological methods in cardinal utility theory, *in* S. K. K. Arrow and P. Suppes (eds), *Mathematical methods in the Social Sciences*, Stanford University Press, 16–26.

[27] Doignon, J.-P. (1987). Threshold representation of multiple semiorders, *SIAM Journal on Algebraic and Discrete Methods* **8**: 77–84.

[28] Doignon, J.-P. (1988). Sur les représentations minimales des semiordres et des ordres d'intervalles, *Mathématiques et Sciences Humaines* **101**: 49–59.

[29] Doignon, J.-P., Ducamp, A. and Falmagne, J.C. (1987). On the separation of two relations by a biorder or a semiorder, *Mathematical Social Sciences* **13**: 1–18.

[30] Doignon, J.-P. and Falmagne, J.C. (1994). Well graded families of relations, *Technical report*, Institute of Mathematical Behavioral Sciences, University of California, Irvine. To appear in Discrete Mathematics.

[31] Doignon, J.-P., Monjardet, B., Roubens, M. and Vincke, Ph. (1986). Biorder families, valued relations and preference modelling, *Journal of Mathematical Psychology* **30**(4): 435–480.

[32] Domotor, Z. and Stelzer, J.H. (1971). Representation of finitely additive semiordered probability structures, *Journal of Mathematical Psychology* **8**: 145–158.

[33] Dubois, D. and Prade, H. (1988). *Possibility theory - An approach to computerized processing of uncertainty*, Plenum Press.

[34] Dushnik, B. and Miller, E. (1941). Partially ordered sets, *American Journal of Mathematics* **63**: 600–610.

[35] Eswaran, K.P. (1975). Faithful representation of a family of sets by a set of intervals, *SIAM Journal on Computing* **4**(1): 56–68.

[36] Falmagne, J.C. and Doignon, J.-P. (1995). Stochastic evolution of rationality, *Technical report*, School of Social Sciences, University of California, Irvine. 27 pp.

[37] Fechner, G.T. (1860). *Elemente der Psychophysik*, Breitkof und Hartel.

[38] Fishburn, P.C. (1970). Intransitive indifference in preference theory: a survey, *Operations Research* **18**(2): 207–228.

[39] Fishburn, P.C. (1974). Lexicographic orders, utilities and decision rules: a survey, *Management Science* **20**(11): 1442–1471.

[40] Fishburn, P.C. (1976). Noncompensatory preferences, *Synthese* **33**: 393–403.

[41] Fishburn, P.C. (1985). *Interval orders and intervals graphs*, Wiley.

[42] Fishburn, P.C. (1991a). Nontransitive additive conjoint measurement, *Journal of Mathematical Psychology* **35**: 1–40.

[43] Fishburn, P.C. (1991b). Nontransitive preferences in decision theory, *Journal of Risk and Uncertainty* **4**: 113–134.

[44] Fishburn, P.C. (1997). Generalizations of semiorders, *Technical report*, AT&T Research, Murray Hill, NJ 07974. To appear in Journal of Mathematical Psychology.

[45] Fishburn, P.C. and Monjardet, B. (1992). Norbert wiener on the theory of measurement (1914, 1915, 1921), *Journal of Mathematical Psychology* **36**: 165–184.

[46] Flament, Cl. (1983). On incomplete preference structures, *Mathematical Social Sciences* **5**: 61–72.

[47] Garey, M.R. and Johnson, D.S. (1979). *Computers and intractability: A Guide to the theory of NP-completeness*, Freeman.

[48] Georgescu–Roegen, N. (1936). The pure theory of consumer's behavior, *Quarterly Journal of Economics* **50**: 545–593.

[49] Gilboa, I. and Lapson, R. (1995). Aggregation of semiorders: intransitive indifference makes a difference, *Journal of Economic Theory* 109–126.

[50] Goldstein, W.M. (1991). Decomposable threshold models, *Journal of Mathematical Psychology* **35**: 64–79.

[51] Golumbic, M.C. and Shamir, R. (1993). Complexity and algorithms for reasoning about time: a graph-theoretic approach, *Journal of the ACM* **40**: 1108–1133.

[52] Goodman, N. (1951). *Structure of appearance*, Harvard University Press.

[53] Guilbaud, G. (1978). Continu expérimental et continu mathématique, *Mathématiques et Sciences Humaines* **62**: 11–34.

[54] Halphen, E. (1955). La notion de vraisemblance, *Publications de L'I.S.U.P* **4**(1): 41–92.

[55] Hansen, E. (1992). *Global optimization using interval analysis*, M. Dekker.

[56] Hansson, B. and Sahlquist, H.A. (1976). A proof technique for social choice with variable electorate, *Journal of Economic Theory* **13**: 193–200.

[57] Herstein, I. and Milnor, J. (1953). An axiomatic approach to measurable utility, *Econometrica* **21**: 291–297.

[58] Hillier, F.S. and Lieberman, G.J. (1990). *Introduction to Operations Research*, 5th edn, McGraw-Hill.

[59] Hodson, F.R., Kendall, D.G. and Tautu, P. (1971). *Mathematics in the Archaeological and Historical Sciences*, Edinburgh University Press.

[60] Hubert, L. (1974). Some applications of graph theory and related non-metric-techniques to problem of approximate seriation: the case of symmetric proximity measures, *The British Journal of Mathematical and Statistical Psychology* **27**(2): 133–135.

[61] Hudry, O. (1989). *Recherche d'ordres médians: complexité, algorithmique et problèmes combinatoires*, PhD thesis, E.N.S.T., Paris.

[62] Jacquet–Lagrèze, E. (1978). Représentation de quasi-ordres et de relations probabilistes transitives sous forme standard et méthodes d'approximation, *Mathématiques et Sciences Humaines* **63**: 5–25.

[63] Kaplan, H. and Shamir, R. (1996). Pathwidth, bandwidth and completion problems to proper interval graphs with small cliques, *SIAM Journal on Computing* **25**(3): 540–561.

[64] Karp, R.M. (1993). Mapping the genome: some combinatorial problems arising in molecular biology, *Proceedings 25th STOC*, ACM Press, 278–285.

[65] Kendall, D.G. (1963). A statistical approach to flinders petrie's sequence dating, *Bulletin of the International Statistical Institute* **40**: 657–680.

[66] Kendall, D.G. (1969a). Incidence matrices, interval graphs and seriation in archaelogy, *Pacific Journal of Mathematics* **28**: 565–570.

[67] Kendall, D.G. (1969b). Some problems and methods in statistical archaelogy, *World Archaeology* **1**: 68–76.

[68] Kendall, D.G. (1971a). Abundance matrices and seriation in archaeology, *Zeitschift für Wahrscheinlichkeitstheorie verw. Geb* **17**: 104–112.

[69] Kendall, D.G. (1971b). A mathematical approach to seriation, *Philosophical Transactions of the Royal Society of London A* **269**: 125–135.

[70] Klir, G. and Bo Yuan (1995). *Fuzzy sets and fuzzy logic; theory and applications*, Prentice Hall.

[71] Kraft, C.H., Pratt, J.W. and Seidenberg, A. (1959). Intuitive probability on a finite sets, *Annals of Mathematical Statistics* **30**: 408–419.

[72] Krantz, D.M., Luce, R.D., Suppes, P. and Tversky, A. (1978). *Foundations of Measurement I*, Academic Press.

[73] Laporte, G. (1987). Solving a family of permutation problems on 0–1 matrices, *RAIRO* **21**(1): 65–85.

[74] Lawler, E. (1976). *Combinatorial optimization: networks and matroids*, Holt, Rinehart and Winston.

[75] Luce, R.D. (1956). Semi-orders and a theory of utility discrimination, *Econometrica* **24**: 178–191.

[76] Luce, R.D. (1973). Three axiom systems for additive semiordered structures, *SIAM Journal of Applied Mathematics* **25**: 41–53.

[77] MacCrimmon, K.R. and Larsson, S. (1979). Utility theory: axioms versus paradoxes, *in* M. Allais and O. Hagen (eds), *Expected Utility hypotheses and the Allais paradox*, Reidel Dordrecht, 333–409.

[78] Marchant, Th. (1996). Valued relations aggregation with the Borda method, *Journal of Multicriteria Decision Analysis* **5**: 127–132.

[79] May, K.O. (1954). Intransitivity, utility and the aggregation of preference patterns, *Econometrica* **22**: 1–13.

[80] Mitas, J. (1994). Minimal representation of semiorders with intervals of same length, *in* V. Bouchitté and M. Morvan (eds), *Orders, algorithms and applications*, number 831 in *Lecture Notes in Computer Science*, Springer, 162–175.

[81] Monjardet, B. (1978). Axiomatiques et propriétés des quasi-ordres, *Mathématiques et Sciences Humaines* **63**: 51–82.

[82] Monjardet, B. (1984). Probabilistic consistency, homogeneous families of relations and linear Λ-relations, *in* E. Degreef and J. van Buggenhaut (eds), *Trends in Mathematical Psychology*, North-Holland, 271–281.

[83] Moore, R. (1966). *Interval Analysis*, Prentice–Hall.

[84] Morrison, H.W. (1962). *Intransitivity of paired comparaison choices*, PhD thesis, University of Michigan.

[85] Nagaraja, R. (1992). Current approaches to long-range physical mapping of the human genome, *in* R. Anand (ed.), *Techniques for the analysis of complex genomes*, Academic Press, 1–18.

[86] Nitzan, S. and Rubinstein, A. (1981). A further characterization of Borda ranking method, *Public choice* **36**: 153–158.

[87] Pawlak, Z. (1991). *Rough Sets. Theoretical aspects of reasoning about data*, Kluwer.

[88] Petrie, W. (1899). Sequences in prehistoric remains, *Journal of the Anthropological Institute* **29**: 295–301.

[89] Pirlot, M. (1990). Minimal representation of a semiorder, *Theory and Decision* **28**: 109–141.

[90] Pirlot, M. (1991). Synthetic description of a semiorder, *Discrete Applied Mathematics* **31**: 299–308.

[91] Pirlot, M. (1996). A common framework for describing some outranking methods. To appear in *Journal of Multicriteria Decision Analysis*.

[92] Pirlot, M. and Vincke, Ph. (1992). Lexicographic aggregation of semiorders, *Journal of Multicriteria Decision Analysis* **1**: 47–58.

[93] Poincaré, H. (1905). *La valeur de la science*, Flammarion, chapter 3.

[94] Rabinovitch, I. (1978). The dimension of semiorders, *Journal of Combinatorial Theory (A)* **25**: 50–61.

[95] Roberts, F.S. (1968). *Representations of indifference relations*, PhD thesis, Departement of Mathematics, Stanford University, Stanford, CA.

[96] Roberts, F.S. (1969). Indifference graphs, *in* F. Harary (ed.), *Proof Techniques in Graph Theory*, Academic Press, 301–310.

[97] Roberts, F.S. (1971a). Homogeneous families of semiorders and the theory of probabilistic consistency, *Journal of Mathematical Psychology* **8**: 248–263.

[98] Roberts, F.S. (1971b). On the compatibility between a graph and a simple order, *Journal of Combinatorial Theory* **11**: 28–38.

[99] Roberts, F.S. (1976). *Discrete Mathematical Models with Applications to Social, Biological and Environmental Problems*, Prentice–Hall.

[100] Roberts, F.S. (1978). *Graph theory and its applications to problems of society*, Vol. 29 of *CBMS-NSF Regional conference Series in Applied Mathematics*, SIAM.

[101] Roberts, F.S. (1979). *Measurement theory, with applications to Decision Making, Utility and the Social Sciences*, Addison-Wesley.

[102] Robinson, W.S. (1951). A method for chronologically ordering archaeological deposits, *American Antiquity* **16**: 293–301.

[103] Rockafellar, R.T. (1972). *Convex Analysis*, Princeton University Press.

[104] Rogers, D.G. (1977). Similarity relations on finite ordered sets, *Journal of Combinatorial Theory (A)* **23**: 88–99.

[105] Roubens, M. and Vincke, Ph. (1984). A definition of partial interval orders, *in* E. Degreef and J. V. Buggenhaut (eds), *Trends in Mathematical Psychology*, North-Holland, 309–316.

[106] Roubens, M. and Vincke, Ph. (1985). *Preference modelling*, number 250 in *Lecture Notes in Economics and Mathematical Systems*, Springer.

[107] Roubens, M. and Vincke, Ph. (1988). Fuzzy possibility graphs and their application to ranking fuzzy numbers, *in* J. Kacprzyk and M. Roubens (eds), *Nonconventional preference relations in decision making*, number 301 in *Lecture Notes in Economics and Mathematical Systems*, Springer, 119–128.

[108] Roy, B. (1968). Classement et choix en présence de points de vue multiples (la méthode Electre), *Revue Française d'Informatique et de Recherche Opérationnelle* **8**: 57–75.

[109] Roy, B. (1969). *Algèbre moderne et théorie des graphes*, Dunod. Volumes I and II.

[110] Roy, B. (1974). Critères multiples et modélisation des préférences: l'apport des relations de surclassement, *Revue d'Economie Politique* **1**: 1–44.

[111] Roy, B. (1985). *Méthodologie Multicritère d'Aide à la Décision*, Economica.

[112] Roy, B. (1993). Decision science or decision-aid science ?, *European Journal of Operational Research* **66**(2): 184–204.

[113] Roy, B. and Bouyssou, D. (1993). *Aide Multicritère à la Décision: Méthodes et cas*, Economica.

[114] Roy, B. and Hugonnard, J.-Chr. (1982). Ranking of suburban line extensions projects on the Paris metro system by a new multicriteria method, *Transportations Research* **16A**: 301–312.

[115] Roy, B. and Vincke, Ph. (1987). Pseudo-orders: definition, properties and numerical representation, *Mathematical Social Sciences* **62**: 263–274.

[116] Scott, D. (1964). Measurement structures and linear inequalities, *Journal of Mathematical Psychology* **1**: 233–247.

[117] Sharp, Jr. H. (1971/72). Enumeration of transitive, step-type relations, *Acta mathematica Academiae Scientiarum Hungarica* **22**: 365–371.

[118] Słowiński, R. (1992). A generalisation of the indiscernibility relation for rough sets analysis of quantitative information, *Rivista di Matematica per le Scienze Economiche e Sociali* **15**(1): 65–78.

[119] Słowiński, R. and Vanderpooten, D. (1995). Similarity relation as a basis for rough approximations, *ICS Research Report 53195*, Warsaw University of Technology.

[120] Suppes, P., Krantz, D.M., Luce, R.D. and Tversky, A. (1989). *Foundations of Measurement II*, Academic Press.

[121] Troxell, D.S. (1995). On properties of unit interval graphs with a perceptual motivation, *Mathematical Social Sciences* **30**: 1–22.

[122] Tversky, A. (1969). Intransitivity of preferences, *Psychological Review* **76**: 31–48.

[123] Vansnick, J.-C. (1986). On the problem of weights in multiple criteria decision making, *European Journal of Operational Research* **24**: 288–294.

[124] Vincke, Ph. (1978). Quasi–ordres généralisés et représentation numérique, *Mathématiques et Sciences Humaines* **62**: 35–60.

[125] Vincke, Ph. (1979). Preferences representation in actuarial sciences, *Scandinavian Actuarial Journal* 223–230.

[126] Vincke, Ph. (1980). Linear utility functions on semiordered mixture spaces, *Econometrica* **3**(48): 771–776.

[127] Vincke, Ph. (1988). (P,Q,I)–preference structures, *in* J. Kacprzyk and M. Roubens (eds), *Nonconventional preference relations in decision making*, number 301 in *Lecture Notes in Economics and Mathematical Systems*, Springer, 72–81.

[128] Vincke, Ph. (1992). *Multicriteria Decision-Aid*, Wiley.

[129] Vind, K. (1991). Independent preferences, *Journal of Mathematical Economics* **20**: 119–135.

[130] Von Neumann, J. and Morgenstern, O. (1947). *Theory of games and economic behavior*, Princeton University Press.

[131] Wiener, N. (1919–1920). A new theory of measurement: a study in the logic of mathematics, *Proceedings of the London Mathematical Society* **19**: 181–205.

[132] Wine, R.L. and Freund, J.E. (1957). On the enumeration of the decisions patterns involving n means, *Annals of Mathematical Statistics* **28**(1): 256–259.

[133] Young, H.P. (1974). An axiomatization of Borda's rule, *Journal of Economic Theory* **9**: 43–52.

[134] Zadeh, L.A. (1965). Fuzzy sets, *Information and Control* **8**: 338–353.

[135] Zadeh, L.A. (1978). Fuzzy sets as a basis for a theory of possibility, *Fuzzy Sets and Systems* **1**: 3–28.

Author Index

Index

THEORY AND DECISION LIBRARY

SERIES B: MATHEMATICAL AND STATISTICAL METHODS
Editor: H. J. Skala, *University of Paderborn, Germany*